Chemistry Transformed:
The Paradigmatic Shift
from Phlogiston
to Oxygen

MODERN SOCIOLOGY:
A Series of Monographs, Treatises, and Texts

Edited by
GERALD M. PLATT

Chemistry Transformed:
The Paradigmatic Shift
from Phlogiston
to Oxygen

H. GILMAN McCANN
University of Vermont

ABLEX PUBLISHING CORPORATON
Norwood, New Jersey

1978

Ablex Publishing Corporation
355 Chestnut Street
Norwood, New Jersey 07648

Library of Congress Cataloging in Publication Data

McCann, H   Gilman, 1942-
   Chemistry transformed.

   (Modern Sociology)
   Includes bibliographical references and index.
   1. Chemistry—History. 2. Science—Social aspects.
I. Title. II. Series.
QD11.M23   540'.9'033   78-19173
ISBN 0-89391-004-X

Printed in the United States of America

Contents

Preface

More than 200 years ago the sequence of events known as the Chemical Revolution began in France. The revolutionary views of Antoine Lavoisier concerning combustion, weight, acidity, and the nature of a chemical element resulted in what is now considered the birth of modern chemistry. In 1785 major French chemists, such as Berthollet and Fourcroy, converted to this new system and by the end of the century the oxygen theory of combustion had become the orthodoxy of the European scientific community. The same period also nurtured the development of chemistry, and science in general, as a professional career and the establishment of the specialized scientific periodical.

The present study, first begun some dozen years ago, treats those events and processes as both an historical case study and as empirical data against which to evaluate the application of my general theory of scientific revolutions. The theory is based primarily upon the highly suggestive ideas in Thomas Kuhn's *The Structure of Scientific Revolutions* (1962) and secondarily upon the view of communication and social control in science developed in the works of Robert Merton and, especially, Warren Hagstrom's *The Scientific Community*. Using a quantitative approach to the social history of science, I hope both to test and refine that theory and to shed new light on history and historical processes.

Obviously, in a project as ambitious as this there are many people who contributed, although I take full responsibility for the results. First I would like to thank Richard Hilbert of the University of Oklahoma, who awakened in me an interest in sociology. My major debt is to Thomas Kuhn, whose work was the inspiration for mine and whose many criticisms and comments were most helpful. I also owe a great deal to Marion Levy, Jr., for theoretical criticism and to Robert Althauser and Kent Smith for instruction in and help with methodology and quantitative analysis. And to Marvin Bressler whose ready wit and kind encouragement kept me going.

I would also like to thank the many students and friends who have aided me, especially G. T. Nygreen at Princeton, Edward Dunphy at the University of New Mexico, and Joe Skudlarek at Princeton, who contributed computer programs, and Craig Palm at the University of Vermont, who helped in data preparation as well as proofing the final copy. Nick Danigelis and Ralph Underhill of the University of Vermont were very helpful in editing the manuscript. I am grateful, too, to the computing centers at Princeton University, the University of New Mexico, and the University of Vermont for free time on their machines and assistance from their staffs.

The real work began in libraries. I owe special thanks to the many helpful librarians of the following libraries: Academy of Natural Sciences Library, Philadelphia; American Museum of Natural History Library, New York; American Philosophical Society Library, Philadelphia; Boston Athenaeum; Boston Public; College of Physicians Library, Philadelphia; Countway, Boston; Engineering Society Library, New York; Library of Congress; National Library of Medicine, Bethesda, Maryland; New York Academy of Medicine Library; New York Public; and the libraries of the following Universities: California at Berkeley, Columbia, Harvard, Johns Hopkins, Massachusetts Institute of Technology, New York, Princeton, and Stanford.

Finally, I would like to thank my wife Audrey. Without her moral support this work would never have been completed.

Burlington, Vermont　　　　　　　　　　　　　　H. GILMAN MCCANN
　　　March 17, 1978

To the memory of my father,
Harold G. McCann,
chemist

Le XVIII siècle fut une époque capitale pour le progrès de la chimie. Placé au début sous l'influence de la théorie du phlogistique, il s'est achevé avec la révélation de phénomènes qui marquerent la naissance de la chimie moderne.

CHARLES BEDEL

Introduction

When Thomas Kuhn's *The Structure of Scientific Revolutions* (1962) was published, it created a stir among philosophers of science. Because of its emphasis on subjective elements in the progress of science, it represented an attack on the prevailing ethos.[1] However, since many of the ideas that aroused the strongest criticism tend to be sociological in nature, they have provided inspiration for sociologists of science, as noted in a recent review of the field (Ben-David & Sullivan, 1975). For the purposes of this study, I have used Kuhn's ideas as a starting point and have accepted them as "true" in order to explore their fruitfulness for a more detailed study of scientific revolutions. I have applied these ideas, together with some hypotheses about social control in science developed in Hagstrom's *The Scientific Community* (1965), to the Chemical Revolution of the late eighteenth century. This revolution was built upon a new theory of combustion and Antoine Lavoisier's new conception of the nature of chemical elements, and it is considered to be the birth of modern chemistry. In addition, I have developed and tested a system of hypotheses about the characteristics of journal publications during this period.

Some of the ideas of Kuhn and Hagstrom are controversial and have been the subject of criticism.[2] My basic strategy, however, has been to suspend judgment,

[1]See, for example, Scheffler (1967) and Lakatos and Musgrave (1970).

[2]For a discussion of the relevant criticism, see Chapter 1.

to assume that these ideas are correct in order to see what kind of theory can be derived from them, and to rely on data to determine how useful this theory is in accounting for the observed phenomena. In a sense, my strategy is based on Kuhn's premise that science is successful in part because it relies on the basic theories used to guide research — which function, in philosophical terms, as inference tickets — and not because it subjects these theories to direct empirical test. Too often, many sociologists criticize "prematurely"; they denigrate theories because of esthetic, metaphysical, or logical flaws rather than applying them to data in order to assess their usefulness.

The theory of scientific revolutions presented here follows the general outline of the ideas formulated by Kuhn, although the specifics are entirely mine. Although I have been inspired and guided by *The Structure of Scientific Revolutions,* I do not claim to be presenting these ideas as Kuhn himself would, but to have developed a theory that includes the biases and predilections of the sociologist rather than the historian. Specifically, sociologists are usually more interested in developing general theories about the ways in which scientists behave, rather than in presenting a detailed historical account of any particular event. The "facts" about the particular events included in this study are primarily intended to specify and assess the theory. However, I firmly believe that we can formulate more fruitful theories about science and scientists if we study the specific substance and histories of the various sciences than if we cling to vague generalities and uncritically accept the myths of science.

I believe this is the first study in which a sociologist has tried to explain and predict processes involving the *content* of a science. Furthermore, I feel that it is one of the few studies in which the actual product of scientists — their professional literature — is used as data.[3] With a few exceptions, such as Fisher (1973) and studies of the spread of innovations, sociologists have dealt with neither the substance nor the products of science, but only its institutional context.

Although I hope these characteristics have made this study important, they have also led to certain problems. Because the theory is a first attempt, based on rather general and untested ideas, the data are rather crude, and the results in some areas may only point up inconsistencies, ambiguities, or even errors. In spite of this, however, the study seems suggestive and should prove fruitful for further research. It is better for our purposes to be clear and wrong than vague and "right." In fact, I expect the questions raised by this study will be at least as important as the answers given; in an exploratory study such as this, one can expect no more.

What I have done is to take the Chemical Revolution as the source of raw data in order to see if the hypotheses suggested by its own literature help to illuminate the progress of this Revolution. Relying primarily on secondary sources for information on the period preceding the crisis precipitated by Lavoisier, I

---

[3]This is true to some extent for Fisher (1966, 1967, 1973) and Merton's early work, reprinted in 1968 and 1970.

have used primary sources — published journal literature — to determine the reactions to Lavoisier's theories and to study the process of their acceptance and dissemination by communities of chemists.

The reason that I have chosen a revolutionary period is based on Kuhn's implication that such a period is sociologically more interesting and revealing; it is during such periods that the goals, norms, and basic values of the scientific community are challenged, and even scientists of the time realize that the solution to problems may involve more than purely logical considerations. Furthermore, as Hagstrom points outs, the latent functions (the ones that are problematical and, therefore, illuminating) of the communication channels come to light during these periods of goal conflict.

I have also chosen this particular revolution because (a) it is intrinsically interesting; (b) the subject matter is not too esoteric for a nonspecialist to become acquainted with in a relatively short period of time; (c) superficially, at least, it exemplifies the points made by Kuhn (he uses it as a major example); and (d) there is much secondary literature to point the way, although historians have not concerned themselves with the sociological implications of their findings. It is my hope that the results of this study will prove as valuable to historians as to sociologists.

I have focused specifically on France and Great Britain because they were the countries in which most of the work leading to the Revolution was carried out,[4] including the findings of such central figures as Lavoisier, Priestley, Cavendish, Black, Kirwan, Berthollet, Fourcroy, and Guyton de Morveau. Although one can argue for an international community of chemists (cf. Langer, 1972) including representatives from Germany, Scandinavia, Italy, Holland, and Russia, as well as France and Britain, it is clear that the latter two stand out. This is not to deny that major results were contributed by such men as Bergman (with his definitive treatise on affinities), Scheele (who some claim as the discoverer of oxygen), or Meyer (an important phlogiston theorist), but the centers of opposition were obviously France (which produced the new oxygen theory of Lavoisier and his associates) and Great Britain (the home of such defenders of phlogiston as Priestley and Kirwan). Therefore, the most appropriate location of the initial test is the chemical literature of these two communities.[5]

[4]Partington, for example, in his monumental history of chemistry (1962) devotes more than 250 pages to French chemistry between 1750 and 1800 (including a chapter on Lavoisier and one on Berthollet and Morveau), over 270 pages to British chemistry from the time of Hales, and only 80 or so pages to Scandinavian and about 50 to German chemistry of the same period. Even the German, Meyer (1906), gives many more pages to French and British chemistry of this time than German and Scandinavian.

[5]Data relative to journal literature (Chapter 3 and Appendix B) and citations (Chapter 6) provide some indication that the historical record needs reevaluation and that Germany might well provide a useful additional source (cf. Hufbauer, 1971). The fact remains, however, that German work was less significant for the Revolution than either French or British.

The events and processes conveniently labeled "The Chemical Revolution" are varied and complex. Hidden among them are many unknowns, many unsolved problems, many questions not yet even asked. I have focused on only a few of the concepts and developments that brought about the shift from the "phlogiston" paradigm, which dominated chemistry in Europe during the middle of the eighteenth century, to the "oxygen" paradigm which, in a sense, we still use today. My treatment centers on the phenomena of combustion and the calcination of metals (what was then called roasting, and is today referred to as oxidation) with secondary consideration given to the concept of affinities. This concept is most closely approximated by the modern notion of "replacement series" and was originally based on the Newtonian conception of matter and the attractive forces between particles. While I have concentrated on these topics, I should note that the Revolution also involved changes in the conceptions of matter, chemical elements, and acidity as well as a theory of heat and change of state (cf. Schofield, 1970; Siegfried & Dobbs, 1968; Smeaton, 1970).

There are three concepts which the reader not versed in eighteenth-century chemistry may need explained: phlogiston, calcination, and affinity. Phlogiston, as will become clear in later chapters, was perceived to be the cause of combustion and related phenomena before the Chemical Revolution took place: those bodies which are combustible contain phlogiston. The more phlogiston they contain, the more combustible they are, so that substances such as oil or coal were viewed as almost pure phlogiston. When hydrogen was discovered, it was thought by many phlogistonists to be phlogiston itself. Phlogiston had some other useful properties as well. Not only were all combustible materials seen as similar and combustible because of the presence of phlogiston, but according to some versions of the theory, which tended to identify phlogiston as sulphur, all acids were viewed as merely modifications of vitriolic (sulfuric) acid. As we shall see, part of the revolution involved a new theory of acidity.

Metals formed a third class of substances united by phlogiston. All metals were alike, that is, metallic, because they all contained the same basic ingredient, phlogiston. When phlogiston was lost through the process of roasting or heating in a flame, a dross, or *calx*, remained, hence the term *calcination*. Thus, two disparate phenomena were united: combustion and calcination or, in modern terms, combustion and oxidation (fast and slow oxidation). These two phenomena are still seen as similar, but our explanation of the process is different; the modern view is the product of the Revolution which we are investigating.

The idea of affinity was akin to our modern notion of replacement series. Oversimplifying for the sake of clarity, we can say that affinity was a manifestation of Newtonianism. Substances had an affinity for each other in much the same way that they had gravitational attraction. It was known that when certain substances were placed in solution with other substances, certain of them tended to combine with one another to the exclusion of others. Substances which had a strong affinity for each other would tend to displace from the solution substances having a lower affinity. Those that combined were said to have an attraction, or "affinity," for one another. Thus, if we were to take a compound of A and B and put them into a solution with C and find that C combines with A to the exclusion of B, then we would say that C has a greater affinity for A than for B and that A has a greater affinity for C than for B.

I will assume the reader is familiar with the concept of "oxygen." While it did not have quite the same meaning for the chemists of the eighteenth century as it does for us today, it is close enough so that the general meaning should be clear. For the less technically inclined reader, I will merely point out that from a modern standpoint, combustion involves the combination of a burning substance (or part of it) with the element oxygen; whereas from the standpoint of the phlogiston paradigm, combustion consisted of the burning substance giving up its phlogiston.

These basic concepts and their relationship will be fully clarified in context. The results and discussions should demonstrate the usefulness of working with a clear theoretical model and basing conclusions on careful, quantitative research. This method will contribute to both the historical record and to the development and refinement of a sociological approach to the understanding of science.

PLAN OF THE BOOK

In Chapter 1, I will develop the general theory of scientific revolution, the general laws, and in Chapter 2 provide the necessary historical detail for operationalizing the theory, resulting in twenty specific hypotheses.

Chapter 3 provides some background material, based heavily on secondary literature, and an examination of the structure of the chemical communities in France and Great Britain. It is here that the testing of the theory begins. Chapters 4 and 5 contain the bulk of the hypothesis testing; Chapter 4 focuses on basic hypotheses about changes during the revolution, and Chapter 5 looks closely at causal models of the process. Chapter 6 concludes the hypothesis testing and develops the conception of community structure further through the use of citation data. In Chapter 7, I present a summary and the general conclusions of the study.

There are also three appendices: Appendix A contains a discussion of the methodology employed, a description of the variables, the samples, and the techniques; Appendix B gives a brief description and comparison of the important journals of the period; and Appendix C presents some supplementary tables, useful for comparing results in different subgroups of the data.

1

A General Theory of
Scientific Revolutions

The starting point for a general theory of scientific revolutions is the work of Thomas Kuhn (1961, 1962, 1970). First we must distinguish revolutionary science from "normal" science. Normal science is the science in which scientists are engaged most of the time — science as presented in textbooks — and the kind of science on which the sociology of science has concentrated. According to Kuhn (1962), normal science is conducted in terms of an accepted "paradigm." The concept of paradigm is a controversial one which includes both a basic theory and a set of accepted research techniques and practices:

> [Paradigm is] a term that relates closely to 'normal' science. By choosing it I mean to suggest that some accepted examples of actual scientific practice — examples which include law, theory, application, and instrumentation together — provide models from which spring particular coherent traditions of scientific research. [p. 10]

This notion of paradigm has received much criticism,[1] and Kuhn has refined it for the second edition of his book (Kuhn, 1970):

> In much of the book the term 'paradigm' is used in two different senses. On the one hand, it stands for the entire constellation of beliefs, values, techniques, and so on shared by the members of a given community. On the other, it denotes one sort of element in that constellation, the concrete puzzle-solutions which, employed as models or examples, can replace explicit rules as a basis for the solution of the remaining puzzles of normal science. [p. 175]

Since it is more appropriate for sociological analysis, our study uses "paradigm" in the first sense.

[1] See especially Lakatos and Musgrave (1970). A variety of Kuhn's ideas are attacked here, mostly on the grounds that they are too "sociological"; that is, they assert that there is a large nonlogical component to science. See also footnote 12, p. 13.

It is important to note that a paradigm is "shared by the members of a given community," and Kuhn has emphasized this aspect of the concept in his 1970 postscript:

> A paradigm governs, in the first instance, not a subject matter but rather a group of practitioners. Any study of paradigm-directed or of paradigm-shattering research must begin by locating the responsible group or groups. [p. 180]

> [A] scientific community consists, on this view, of the practitioners of a scientific specialty... The members of a scientific community see themselves and are seen by others as the men uniquely responsible for the pursuit of a set of shared goals. [p. 177]

Recent work in the sociology of science has also placed heavy emphasis on the concept of community.[2] Warren Hagstrom's appropriately titled *The Scientific Community* is typical in this respect and provides the most extended discussion of community. His central concern is the patterns and implications of control in scientific communities. He notes that scientists who are highly committed to the goals, values, and methods of science are produced by a "highly selective system of recruitment" (Hagstrom, 1965, p. 11), as well as by the process of socialization into a scientific community:

> The context of the texts, lectures, and laboratory work presented in the course of higher education in the sciences integrates the general norms and values of science with a specific set of beliefs and techniques. [p. 11]

However, he feels that exclusive reliance on these factors leads to an overly individualistic account of science. By nature, scientific communities do not just happen to "hang together." As in any other social group, there are mechanisms of social control:

> Not only is the individualistic view directly controverted by obvious facts about the scientific community, but there is every theoretical reason to expect this to be so. First, the autonomy of the scientific community cannot be taken for granted, it must be maintained by internal social controls among other things. Without them, scientists would tend to respond more readily to the goals and standards of nonscientists. [p. 12]

Hagstrom's model of control is based upon "exchange theory," in which social structure is built from interpersonal exchanges.[3] He begins by observing that science is autonomous and suggests that "the organization of science consists of an exchange of social recognition for information" (p. 13). Recognition, rather than money or power, is the basic motivating force in science, and it

---

[2] This concept, of course, has a longer history in the sociology of science: Merton's early work (reprinted in Merton, 1968, pp. 628-681, 1970) uses the notion to some extent, and Znaniecki's little-used *Social Role of the Man of Knowledge* (1940) makes it a central concern.

[3] While a step away from the self-motivating view he is criticizing, Hagstrom's discussion of recognition (and competition) is still more individualistic than structural.

results, according to Hagstrom, from the "gift" nature of the information:

> Manuscripts submitted to scientific periodicals are often called 'contributions,' and they are, in fact, gifts. [p. 12]

> In general, the acceptance of a gift by an individual or a community implies a kind of recognition of the status of the donor and the existence of certain kinds of reciprocal rights. . . In science, the acceptance by scientific journals of contributed manuscripts establishes the donor's status as a scientist — indeed, status as a scientist can be achieved only by such gift-giving — and it assures him of prestige within the scientific community. [p. 13]

Other sociologists (e.g., Crane, 1965; Storer, 1966; Cole & Cole, 1967) agree that recognition is the prime motivating force in science, but do not view it as a gift exchange — and it remains to be seen whether this is indeed the most fruitful conception. Merton, on the other hand, sees recognition not as a goal but rather as a sign that the scientist has done his job:

> Scientific inquiry, like human action generally, stems from a variety and amalgam of motives in which the passion for creating new knowledge is supported by the passion for recognition by peers and the derivative competition for place. [1969, p. 17]

> Recognition for originality becomes socially validated testimony that one has successfully lived up to the most exacting requirements of one's role as a scientist. The self-image of the individual scientist will also depend greatly on the appraisals by his scientific peers of the extent to which he has lived up to this exacting and crucially important aspect of his role. [1957, p. 640][4]

Thus, Merton feels that the desire for recognition is normally subordinate to the desire to promote knowledge by making *original*[5] contributions to the body of scientific lore. Furthermore, on the basis of his studies of multiple discoveries and resistance by scientists to their study (Merton, 1961, 1963b), as well as his work on priorities, he feels that scientists are ambivalent about the desire for recognition (see especially Merton 1963a). This motive, which, along with desires for money or position, he calls "extrinsic," is suspect:

> There is, nevertheless, a germ of psychological truth in the suspicion enveloping the drive for recognition in science. Any extrinsic reward — fame, money, position — is morally ambiguous and potentially subversive of culturally esteemed values. For as rewards are meted out, they can displace the original motive: concern with recognition can displace concern with advancing knowledge. [1969, pp. 17-18]

[4]Here, as elsewhere, Merton is sensitive to analogues suggested by Weber's treatment of the Protestant Ethic. Science is a capricious and unpredictable God, and recognition by peers is one sign that the scientist is, so to speak, among the "elect" who are the beneficiaries of irresistible "grace."

[5]We assume that by "originality" Merton means a novel contribution to a paradigm, the solution to a puzzle. Originality per se is not what is desired or valued by the scientific community. Especially in a period of normal science, highly original work is likely to be viewed with scepticism. On the other hand, there is a desire for priority: recognition is given to the first scientist who contributes an idea that can be incorporated into the paradigm, or in Kuhn's terms, to the scientist who first solves a puzzle. Even in the case of a revolution recognition is not given for pure originality, but for the contribution of a new useful paradigm.

While it is clear that most scientists are motivated to advance knowledge — not the same thing as being original — it is not clear why recognition is considered by Merton to be an extrinsic reward. Is is an integral part of the institution of science, as Nobel prizes and the practice of eponymy show. Rewards may be abused, but that does not make them extrinsic — indeed, it is hard to conceive of social organization without a reward system. Moreover, the data presented in Chapter 3 suggest that a community of scientists cannot flourish on altruistic motives alone.[6] Although science may be seen as a profession, incorporating a sense of a "calling" and social responsibility, it, like other forms of organized human activity, must rely on a system of rewards to attract, motivate, and control its practitioners.

The desire for recognition, whatever its origins, tends to encourage the communication of information from one scientist to another; this can take the form of letters or conversation, so-called "informal" channels of communication. However, due to the large number of scientists and the stress on "public" knowledge, there are also "formal" channels, notably the publication of monographs and journal articles. We find, therefore, that the system of social control is intimately bound to the system of communication.

Hagstrom (1965) distinguishes two forms of recognition, coinciding with the two forms of communication: " 'Institutional recognition' is given in formal channels of communication in science, whereas interpersonal approval and esteem, or 'elementary recognition,' is given in direct communication" (p. 23). This study focuses on formal channels, and in particular, journal publications. As Hagstrom notes:

> Formal communication in the sciences is primarily carried on in articles appearing in scientific journals. . . Recognition of other scientists for their contributions is covered in the same channel of communication. It usually takes the form of either footnote citations to specific articles or a section on 'acknowledgments' to previously published work. [p. 23]

Because of the importance of citations to science and scientists, an importance underscored by the invention of the *Science Citation Index* and, more recently, the *Social Science Citation Index,* they provide a useful tool for investigating the structure of science (see Garvey et al., 1970). One such use was first mentioned by Kuhn (1962): "A shift in the distribution of the technical literature cited in the footnotes to research reports ought to be studied as a possible index to the occurrence of revolutions." (p. xi) A second use, which has become common in recent sociology of science (see especially Crane, 1969) and which is in a sense a generalization of the first, is that of isolating scientific communities by means of "formal and informal communication networks including those discovered in

---

[6]The evidence indicates that a major difference between British and French science in the eighteenth century was the relative absence of a reward system for British scientists, leading to lower productivity for them although one cannot doubt their desire to advance knowledge.

correspondence and in the linkages among citations" (Kuhn, 1970, p. 178). The third use, probably the most widespread, is to measure prestige (see Bayer & Folger, 1966; Cole & Cole, 1967; Crane, 1965; Gaston, 1970; Price, 1970; Zuckerman & Merton, 1971). I will employ all three uses in this study.[7]

With a paradigm, developed and learned through this communication system, to guide him, the scientist engages primarily in "puzzle-solving." The paradigm suggests the choice of problems for study, supplies the methods for attacking these problems, and provides expected outcomes. During this period the paradigm is unquestioned, that is, the scientist does not, as is popularly supposed,[8] subject the community's basic ideas to experiential test, but assumes them to be true and develops them further, pushing them into new areas and solving either new or old puzzles with them. Data which appear to be counter to the theory are not seen as counterexamples or refutations. Instead, the scientist assumes that he can solve the "puzzle" or "anomaly" by fitting it into the paradigm:

> [Normal science is] a strenuous and devoted attempt to force nature into the conceptual boxes supplied by professional education. . . [an attempt which] is predicated on the assumption that the scientific community knows what the world is like. [Kuhn, 1962, p. 5]

A good example of such an attempt, presented by Kuhn (1962, p. 81), relates to problems of applying Newton's law of gravitation to the orbit of the moon, a problem or puzzle which engaged the attention of the best mathematical physicists of the eighteenth century. In a sense, then, the laws contained in the paradigm function as "inference tickets," or rules of inference, rather than as testable hypotheses.[9] We shall see that this same kind of attempt characterized eighteenth-century chemistry in the period before the Chemical Revolution.

There are times, however, when one or more scientists see an anomaly as incorrigible. Then a crisis arises in the scientific community for which the paradigm is relevant, and a dispute develops between those who see the anomaly as a counterexample and those who do not:

> Normal science repeatedly goes astray. And when it does — when, that is, the profession can no longer evade anomalies that subvert the existing tradition of scientific practice — then begin the extraordinary investigations that lead the profession at last to a new set of commitments, a new basis for the practice of science. [Kuhn, 1962, p. 6]

For the dispute to become serious, it is necessary for the opponents of the old paradigm to propose one of their own: "Once it has achieved the status of a

---

[7]There has been some criticism of the general view of recognition presented here (Gustin, 1973, pp. 1120ff.) and of the use of citations as measures of prestige (Blume & Sinclair, 1973). Nevertheless, such a view and such use seem at least good first approximations.

[8]An idea developed by Karl Popper, for example, in Popper (1959, Chapters 1-4). Compare Kuhn (1970, p. 146).

[9]For a similar view, see Nagel, 1961, p. 67.

paradigm, a scientific theory is declared invalid only if an alternate candidate is available to take its place" (Kuhn, 1962, p. 77). It is at this stage that those laws or propositions which are called into question become subject to empirical testing.

Furthermore, it is at this point that measurement or quantification becomes most important. Kuhn (1961) has presented a discussion of the functions of measurement. Although this discussion is intended to describe modern physical science, we will extrapolate from it and assume that the conclusions are also applicable to more primitive sciences, such as eighteenth century chemistry. Two of the major functions of measurement, the discovery of anomalies and the confirmation of theories, become salient during periods of crisis and revolution (Kuhn, 1961, pp. 165-178). As long as discrepancies remain qualitative, it is relatively easy to ignore them or to incorporate them into existing theory. However, it is not so easy to ignore quantitative anomalies:

> When measurement departs from theory, it is likely to yield mere numbers, and their very neutrality makes them particularly sterile as a source of remedial suggestions. But numbers register the departure from theory with an authority and finesse that no qualitative technique can duplicate. [p. 180] [10]

Further:

> No crisis is . . . so hard to suppress as one that derives from a quantitave anomaly that has resisted all the usual efforts at reconciliation. [p. 182]

Once measurement helps precipitate the crisis and a new theory is proposed, the decision between the two theories is usually related to their relative success in explaining the quantitative evidence. It is only at this point that Kuhn considers true confirmation to take place:

> Furthermore, it is for this function — aid in the choice between theories — and for it alone, that we must reserve the word 'confirmation.' We must, that is, if 'confirmation' is intended to denote a procedure anything like what scientists ever do. [p. 184]

Once the dispute is clear, whether quantitative in nature or not, the defenders of the old paradigm will invent any number of ad hoc hypotheses to protect their theory. These hypotheses are ad hoc only in retrospect, of course; if the crisis is resolved without revolution or acceptance of a new theory, they mesh with the rest of the theory as good examples of puzzle-solving. The faithful do not see the anomalies as counterexamples.

The doubters, on the other hand, do not attack the old theory piecemeal or solely in terms of empirical failures; they attack with another theory, a new world view. Thus, we can view a shift of support from one theory to another not merely as a quantitative decision as to which of two formulations fits the facts better, but as a decisive change in the way the world is conceived: a "gestalt

---

[10]The numbers are neutral, that is, in that they do not suggest a new theory. They are not neutral in terms of casting doubt upon the old one.

switch" (Kuhn, 1962, pp. 110-114). In this light, we see that theories are not so much falsified as replaced:

> The act of judgment that leads scientists to reject a previously accepted theory is always based upon more than a comparison of that theory with the world. The decision to reject one paradigm is always simultaneously the decision to accept another, and the judgment leading to that decision involves the comparison of both paradigms with nature and with each other. [p. 77]

Finally, the judgment also involves the status and influence of the scientists involved.

It is important to emphasize this notion of judgment because it provides a point of access for sociological considerations. Since there is, philosophically, no known way of deciding to what extent empirical evidence can verify or falsify a given theory,[11] it is clear that nonlogical, even "unscientific," factors have a bearing on the decision. One such factor, commonly recognized, is simplicity; the simpler of two theories which explain the same data is the better choice. However, this notion is rather subjective: simplicity cannot be defined or decided on purely logical grounds. It involves the perception of the scientists involved, and that perception is, to a great extent, socially and psychologically determined. To complicate matters further, the two competing theories do not explain the same "facts" or data, and proponents of each can and will claim that their theory is the simpler.

Other examples of nonlogical factors are metaphysical views, philosophical positions, ethnocentrism and/or nationalism, and the social characteristics of the scientific community. Indeed, these other factors seem to play an important part at all stages of scientific inquiry, including both the "context of discovery" and the "context of justification."[12] Whatever the ultimate resolution of these problems by philosophers of science, the lack of purely logical criteria for choice means that social factors are likely to be crucial in the historical development of science.

---

[11]See (Hempel, 1964, pp. 3-51) for a discussion of some of the problems surrounding this goal of philosophers of science. Note especially his conclusion:

> Thus the search for purely syntactical criteria of qualitative or quantitative confirmation presupposes that the hypotheses in question are formulated in terms that permit projection; and such terms cannot be singled out by syntactical means alone. Indeed, the notion of entrenchment that Goodman uses for this purpose is clearly *pragmatic* in character. [p. 51, emphasis added]

[12]The "context of discovery" refers to the creative act of inventing hypotheses, which many philosophers have long felt is not amenable to purely logical analysis. The "context of justification" refers to the question of confirming or disconfirming a given theory, and it is commonly felt by philosophers of science to be so amenable. See (Scheffler, 1967, pp. 14-18) for a long discussion of this issue and a concern for the kind of threat which Kuhn's work represents.

In the case of a successful revolution, the paradigm dispute is resolved in favor of a new theory, which consequently guides the scientific community toward a new period of normal science, including a new set of puzzles to be worked out with the confidence that they are indeed solvable. Not only are they assumed to be solvable, but failure of a scientist to solve them reflects his own inadequacy and not the paradigm's (Kuhn, 1961, p. 171).

We have seen that a revolution begins with a crisis period[13] created by the repeated failure of a given paradigm to provide solutions to various anomalies which have arisen. One or more new theories are proposed by those who perceive this crisis and a dispute develops between the supporters of the old paradigm and those of the new candidate. This may result in a debate over several different models, as different versions of both major paradigms are proposed to solve the problems raised: the hypotheses that prove unsuccessful will be viewed as ad hoc, and the successful ones will become part of the accepted paradigm.[14] In the case of a successful revolution, the old paradigm is replaced; the act of replacement does not hinge only on purely logical arguments, but also on the persuasiveness of the scientists proposing the new paradigm. This "persuasiveness" involves appeals to the data, to logic, and to other values of the scientific community.

Finally, there are often those scientists who, because of their lengthy commitment to and success with the old paradigm, are never convinced (and there can be no purely logical grounds on which to convince them) that the new paradigm is superior or more "correct."

These ideas provide the basis for a set of general hypotheses applicable to any scientific revolution, and not just to the one under consideration.

The period of normal science is characterized by puzzle-solving: by such activities as articulating the paradigm, applying it to new areas, working out physical constants, and so forth (Kuhn, 1962, pp. 25-32). Since the paradigm is unquestioned, there is little doubt about the interpretation of the "facts." Many scientists are able to do most of their work without ever worrying about basic theoretical questions, yet some — the theoreticians — will study the theoretical questions, and will try to apply the paradigm to new areas. It is they who will begin the attempt to fit new phenomena, as well as the loose ends of old phenomena, into the appropriate conceptualization.

From such considerations, we can assume that an appropriate or characteristic amount of theoretical work will be done during a period of normal science.[15] That is, in the various publications of the period there will be a certain percentage of articles (fluctuating about some characteristic value) that are highly theoreti-

[13]Kuhn (1970, p. 181) admits that a crisis is not a *necessary* prelude to a revolution.

[14]Kuhn discusses this issue in (Kuhn, 1962, p. 61).

[15]This is an assumption and will not be tested — nor can it be in any individual case.

cal.[16] For similar reasons we shall assume that there is a characteristic amount of quantitative work being published during a period of normal science. A field may take one of two quantitative forms: (1) it may be generally qualitative (as in the period under study), with its characteristic level seen to be the percentage of work that is quantitative; (2) it may be generally quantitative, with its characteristic level seen to be the level of quantitative precision. For our purposes, this study will deal only with the first form.

Whatever the characteristic levels of theory and quantification may be, we hypothesize that they will rise during a period of crisis and/or revolution. In the case of theory, for example, more people will work on a higher theoretical level than during the prior period of normal science, and experimental and applied scientists will become more aware of their theoretical beliefs and attempt to advance them. In short, there will be greater use of basic and general theory in the literature. Likewise, in a revolution involving quantification, such as the one under consideration, papers will become more quantitative in nature, and the proportion of purely qualitative papers will decline. In both cases, the rise is due to the nature of anomalies and the threat they represent for the community.

Anomalies are theoretical by definition, and once they are taken seriously, a crisis is created. Since the crisis is created by theoretical anomaly, attempts to solve it will be theoretical, at least initially. More effort must be put into solving the theoretical problem, since it represents a threat to the paradigm and thus to the social structure evolving from that paradigm. To the extent, therefore, that the crisis dominates a discipline, the hypothesis follows immediately.

Once a new paradigm has been invented to solve the problem(s), the situation becomes even more critical. A new paradigm is "a new basis for the practice of science" (p. 11), and its invention is an attack on the existing social order; new values may result in the reordering of status. Since an attack on a paradigm is an attack on a community's basis for existence, each member of the community must react vigorously: it becomes a matter of survival of the intellectual community, and most scientists take such survival seriously. Kuhn (1962) makes a similar point:

> Like the choice between competing political institutions, that between competing paradigms proves to be a choice between incompatible modes of community life. . . . Each group uses its own paradigm to argue in that paradigm's defense. [p. 93][17]

Because the basis of every paradigm is a theory, the dispute will have a large theoretical component.

[16]If we think of each article as being on a continuum ranging from the most concrete description to the most abstract and general theory, then we are postulating a characteristic mean value for these articles.

[17]While Kuhn's focus is on the circularity of argument, the implications are similar.

Furthermore, since the revolution is primarily one of theories (and, *consequently,* a change in methods, world view, exemplars, and so forth), the shift to a new paradigm occurs earlier among the most theoretical papers. Even in the case in which the shift is one of methodology, it will be concurrent with or subsequent to a shift in theoretical thinking at some level. In fact, we could hardly say that a paradigm shift had occurred, or even that a crisis existed, if it had not appeared in theoretical work.

In addition to this definitional consideration, there is a second reason to believe that the shift appears earlier in theoretical papers. Whatever anomalies exist, their implications are strongest for theoretical assertions. Even though anomalies are generated by experimental failures, they are not anomalies until they have become serious theoretical problems. Furthermore, it is possible for experimenters to continue doing their work even when they cannot account for their results: they can ignore theoretical implications and concerns. Even though "observational" terms are theory-laden, they are less so than pure theoretical terms, and concrete "facts" may be agreed upon by both sides, with varying interpretations of their import. This means that nontheorists, the "fact-gatherers," can ignore the dispute and stay with the old paradigm longer than the theorists (cf. Kuhn, 1962, pp. 121-122).

In more concrete terms: in the case of a scientific revolution we may expect an initially high level of theory among the challengers, since their point of attack is a theoretical one, and a subsequent increase in the theoretical level of articles published by the old paradigm group.

The argument is similar for quantitative work. There would probably be an increase in the level of quantitative work just prior to a crisis, precipitating it. Should the revolution be in any but a purely qualitative field, as it is in this case, the papers of the revolutionary group will be generally quantitative. The defenders of the old paradigm are likely to feel constrained to reply in quantitative terms, although there is no *necessity* for them to engage in quantitative work and arguments. They may, and often do, deny that such considerations are relevant or important. However, given the prevailing ethos of science (which can be traced at least to Newton's time), with its emphasis on measurement and quantity, the pressure is in the quantitative direction: the groups which can make an appeal to quantitative arguments have an advantage over those who cannot. However, if the new paradigm is seen as superior in quantitative arguments, but not necessarily in qualitative ones, the defenders of the old paradigm will increasingly resort to purely qualitative arguments.

Thus, we have discovered some contingencies which are hard to predict. The exact course of quantitative work among the supporters of the old paradigm will depend on particular historical events.[18] In any case, even if the members of the old paradigm group do not involve themselves in quantification, the revolution-

---

[18]In the case under study, most of the defenders avoided quantitative work, especially after the powerful quantitative attack by Lavoisier.

aries will stress this aspect of their work as much as possible. With the success of a revolution, the increased frequency of and dependence upon quantitative work must become institutionalized. Therefore, the level of quantitative work will rise independently of what the supporters of the old paradigm do.

As in the case of theoretical papers, we hypothesize that the shift to the new paradigm will occur first among the quantitative articles. Because quantitative anomalies are the most acute, those who do quantitative work are likely to experience the crisis earliest and most strongly and, therefore, be the first to switch paradigms. Furthermore, once they have converted, they will show a strong tendency to use measurement in every way possible to press their fight. Thus, there should be a very high level of quantitative work among early converts.[19]

Specifically, then, there will be a high level of quantitative work among the revolutionary group, the old paradigm supporters will exhibit either an increase or decrease in their level of quantitative work, and eventually quantitative work will become the norm.

The increase in quantitative work will be most pronounced among theoretical papers. This is where the dispute is most severe: here the anomalies are most noticeable, and quantification sharpens the disparities, so the revolutionaries press their quantitative attack most strongly in this area. If the defenders do choose to reply in quantitative terms, they will do so primarily in their most theoretical work, where it has the greatest effect. Therefore, as the level of theory increases, the amount of quantitative work should likewise increase. However, although the level of theoretical work will decrease after the revolution (albeit slowly, for there are many accompanying problems to solve) to some level characteristic of the new paradigm, the level of quantitative work will not: measurement gains therefore will not be lost.[20]

Accompanying these relative increases, there will be an increase in both the productivity and the number of scientists in the field, an increase greater than that due to the long-term growth rate of science (Price, 1963, Chapter 1) and, perhaps, at the expense of other sciences. The basic cause of this increase is competition for recognition: the first scientist to make a discovery or invent a theory is given credit for it and everyone else loses almost all claims. Merton's studies of

[19]If, on the other hand, the revolution is not successful and the old paradigm survives, the strongest support for the old paradigm should occur among the quantitative papers. This is because the *ad hoc* hypotheses have held up and the anomalies have been explained – the old paradigm has been able to incorporate the quantified data.

[20]At the conclusion of the dispute, quantitative methods will have been institutionalized for at least the problems which are solved by the new paradigm. Thus, we can expect a greater frequency of quantitative work than before the revolution. However, although those areas which have become quantified continue to be dealt with quantitatively, there may be a decline in the overall proportion of quantitative articles in the entire discipline, because new problems may arise which the new paradigm was not created to solve and which are not immediately amenable to quantitative techniques.

priority disputes and multiple discoveries (see especially Merton, 1957, 1961, 1969) provide strong evidence on this point and show that competition for priority can be traced to at least the seventeenth century (see also Hahn, 1971, p. 27).[21]

A field in a state of crisis or revolution will be more attractive to those scientists only marginally connected with it (that is, either those in related fields or those undergoing training), than areas in a "normal" state. Since basic assumptions, both theoretical and methodological, have been called into question, the time is ripe for new ideas. Scientists in related disciplines may feel that their models and methods could be useful in such a field; younger scientists may think that such a field offers them greater opportunity, since the authorities of the past (the experts in the old paradigm) would have no *technical* advantage over them, though they may well have certain nonrational advantages (they are still respected names).

Not only are there no "experts" in the new paradigm nor, at the crisis stage, workers well equipped to solve the persistent anomalies, but, once the new paradigm has been proposed, the tradition to be learned is minimal. Fundamental discoveries are made more easily and result in a shorter wait for significant recognition. In addition to these practical advantages, there is likely to be the pure excitement value of a field undergoing revolution.

However, the situation is not quite this simple. If there is a prolonged and acutely felt crisis, much discouragement for both old and new practitioners may well exist.[22] Kuhn (1962, pp. 83-84) gives some examples of this. Nevertheless, the positive factors would likely outweigh the negative factors: the potential for great gain and the relatively little risk involved are stronger than the frustration that may be experienced (see Hagstrom, 1965, pp. 83-84, for a discussion of risk-taking and Fisher, 1973, pp. 1112-1114, for an example from mathematics). The attraction is even stronger when a crisis passes into a full blown paradigm dispute, for then the potential for frustration is decreased and the excitement increased.

A corollary of this argument is that we expect the supporters of the new paradigm to be predominantly from the younger generation of scientists: it is they who would have less commitment to the old paradigm and more to gain from

---

[21]Hagstrom has a long discussion in (Hagstrom, 1965, Chapter 2) of the signs and consequences of competition and the problems of being anticipated. Compare Collins (1968) and Gaston (1971).

[22]Another complication, which might introduce serious measurement problems, is the question of how great a time lag exists between interest in a field and trained professionalism. In the case of eighteenth century chemistry I doubt that this is a serious problem: there is evidence from the careers of Morveau (lawyer turned chemist) and Lavoisier (geologist turned chemist) that relevant training and knowledge could be acquired in a very short time period – perhaps as little as two years. Furthermore, since many chemists at this time were originally trained as doctors (chemistry was hardly a distinct career until near the end of the century and then only in few places), expertise and productivity in chemistry could take even less time (cf. Smeaton, 1954b, 1961).

adopting the new one (cf. Cole, 1970). As Kuhn (1962) states:

> Almost always the men who achieve these fundamental inventions of a new paradigm
> have been either very young or very new to the field whose paradigm they change. . .
> these are the men who, being little committed by prior practice to the traditional
> rules of normal science, are particularly likely to see that those rules no longer define
> a playable game and to conceive another set that can replace them. [pp. 89-90]

Not only does the number of scientists increase, but their average productivity
(articles per year, for example) also increases. Since logical argument alone is not
enough to demonstrate the superiority of any one paradigm, it is necessary to
overwhelm opponents with persuasion and masses of "evidence." It is in this
situation that the same experiment may be performed many times by members
of both groups, each also complaining about the methodology of the other.[23]

In addition to participating in both theoretical and experimental debates, it is
likely that many of the converts to the new system will repeat some of their
earlier work, using the new explanation to demonstrate their faith and convert
others.[24] All these factors therefore lead to an increase in both the number and
the productivity of scientists.

Now, continuing the gestalt switch analogy, scientists perceive the world
through one paradigm or the other and cannot look "in between." Since there is
much evidence to consider and since each paradigm will fit some parts of it better
than the other (different questions are relevant for different paradigms — one
reason why resistance may be maintained for a long time), there may well be
some period of doubt before any individual scientist knows which paradigm will
prove most useful. However, he is likely to remain loyal to the old paradigm
until the evidence is convincing and then switch to the new one, with little public neutrality.[25] The net result is that there should be few neutral papers; that is,
few papers that expressly state that both theories are adequate with neither
having an edge.[26]

[23]In the case of the Chemical Revolution, for example, Hufbauer (1971, pp. 144-176)
details numerous repetitions of the reduction of mercury calx (mercuric oxide) by both
camps in Germany from 1789 through 1793.

[24]Berthollet, for example repeated much of his earlier work after his conversion to the
new theory in 1785 (see McCann, 1966).

[25]Loyalty to the bitter end should be especially characteristic of prominent scientists,
for they have more to lose and less to gain by converting. However, private doubt and public
faith would be very hard to measure without interviews or private correspondence, so we
cannot test it with the data at hand.

[26]The notion of "neutrality" is problematic: there may be areas of inquiry for which
neither paradigm is adequate (or even relevant, in which case one could say that the area fell
outside the purview of the paradigm even though a subscriber to the paradigm might do research in such an area) or in which one could not distinguish the application of one paradigm from the other. Papers for which one could not decide which paradigm was supported
could be called neutral, but here we restrict this term to papers which expressly mention
neutrality vis-a-vis the competing paradigms (or paradigm candidates).

This provides us with at least one crucial test between the view of science being presented here and the alternative conjecture-refutation view with its emphasis on the decisiveness of data. Assuming the latter view, we would expect the two following patterns: (1) many examples of scientists switching back and forth between competing theories as additional bits of evidence came to light favoring one or the other; and (2) many neutral articles, due to the many instances when some evidence seemed to point one way while other evidence pointed the other (we have already learned that neither paradigm is perfect) so that there are no decisive *evidential* reasons for preferring either.[27]

Finally, let us consider the effect of the revolution on recognition. Clearly, the success of a paradigm implies the success of its proponents. In terms of the measures used here, this means that the predominance of articles using the new paradigm (success of the paradigm) would be accompanied by the predominance of citations to such articles and/or their authors (success of the proponents). Thus, in the case of a revolution, the proportion of articles using the new paradigm will increase from none to all or nearly all, and the proportion of citations to members of the new paradigm group will likewise increase. There will remain, of course, those who are steadfastly unconverted, and there may be references for some time to the heroes of the past or to those who, using the old paradigm, produced some data useful for the new one. However, the unconverted die out and such references to them become increasingly rare. Consequently, if we measure the average age of the citations (age being the time between the publication of the citation and the publication of the piece to which it refers), the age will decrease during the revolution. Older work will gradually become less relevant.[28]

Although there are many ways of conceptualizing and measuring scientific prestige (for example, the election to honorific societies, the receiving of honors, or the mention in texts and histories), I will use the number of references to a scientist's works as the primary indicator. Therefore, the proportion of references to proponents of one or the other of the competing paradigms (and there may be more than two) is an indication of the relative prestige of the groups involved, as well as of the standing of individual members. It is also an indication of the relative prestige or dominance of the paradigms; in a sense it is a better measure of dominance than the proportion of articles using the paradigm. Although a resister can publish many articles even *after* the new paradigm has taken over the field (and some of these resisters may be highly prolific), the resister's work has less chance of being cited.

[27]To anticipate, there is very little evidence of vacillation (see the quotation on Beddoes, p. 35). Further, there are examples of men believing theories in spite of their own contrary evidence. Berthollet is a good example: once he had switched to the new paradigm, he remained convinced even though he had good counterevidence (see McCann, 1966).

[28]See, for example, MacRae (1969), and Price (1970). Unfortunately, measurement of "age" of citations proved impossible for the eighteenth century as citations seldom mentioned article or book titles, or included dates.

It is important to point out that I am referring to citations in professional journals. Authors of popularizations and, especially, textbooks are likely to refer to past heroes of the discipline long after their work is outdated. It is, perhaps, a sign of the immaturity of a scientific discipline when heroes of the past are mentioned frequently in its journal literature: there is no successful paradigm to rule them out. When this occurs, it will probably be in a field caught up in a pre- or multiparadigm stage (Kuhn, 1962, pp. 16-20), in which claims to truth often rely heavily on authority figures. A "scholar's" familiarity with authorities is as important — or more so — as his familiarity with empirical evidence: without a paradigm, there are no facts.

Applying the points I have made to the case at hand, in terms relevant for the Chemical Revolution I have hypothesized the following:

Assumption: There are characteristic levels of theoretical and quantitative work associated with the initial and final periods of normal science.

1. The level of theoretical work will rise as the revolution develops. The rise is composed of (a) an increase in the level of theoretical work among followers of the given paradigm and (b) an initially high level of theoretical work by the followers of the new paradigm, followed by a decline once the success of the new paradigm is assured.

2. The shift to the new paradigm will occur earlier among theoretical papers than among others.

3. The level of quantitative work will rise as the revolution develops. The rise is composed of (a) an increase, possibly followed by a decline, in the level of quantitative work among the followers of the given paradigm and (b) an initially high level of quantitative work among supporters of the new paradigm, possibly followed by a decline as the nnew paradigm succeeds and new problems come to light.

4. The shift to the new paradigm will occur earlier among quantitative papers than among others.

5. The rise in the level of quantitative work will be most pronounced among theoretical papers.

6. There will be an increase in the number of authors as the revolution develops.

7. There will be an increase in the productivity of authors as the revolution progresses.

8. The shift to the new paradigm will occur earlier among papers of younger authors than among papers of older ones.

9. The supporters of the new paradigm will be younger than the defenders of the old one.

10. There will be few neutral papers.

11. The proportion of citations to authors supporting the new paradigm will increase during the revolution.

2

History of the Chemical Revolution

In this chapter I will present a brief history of the Chemical Revolution in order to construct a general framework of what took place and to make the book's hypotheses specific in reference to time, place, and subject matter. After considering the background for the development of chemistry in the late eighteenth century, the chapter will be divided into sections roughly coinciding with the periods of normal science, crisis, paradigm dispute, and resolution of the conflict. I will then conclude with a statement of the system of hypotheses.

BACKGROUND

The major early influence on eighteenth century chemistry was the triumph of Newton's mechanical philosophy (Boas, 1959, p. 499). As chemistry slowly became a science, two views of matter were developed from this philosophy. Although there is not complete agreement among historians as to the details of each view, we find a general distinction between those who thought in terms of (relatively observable) chemical principles and those who thought in terms of ultimate physical particles or elements (Boas, 1959):

> On the one hand are those chemists who combined a belief in the existence of small particles as the ultimate entities necessary for the structure of matter with a strong emphasis on chemical principles or elements as the real instruments of chemical change. [On the other hand] were the chemists who so firmly incorporated the notion of a particulate structure of matter into their thinking that they automatically spoke and thought of chemical composition and chemical reaction as involving the arrangement and rearrangement of chemical units in the form of particles. [pp. 500-501]

Schofield (1970, pp. 95-96) finds a similar distinction between "materialists" and "mechanists," and Crosland (1963, pp. 373-393) distinguishes between the followers of Stahl and of Boerhaave.[1]

[1] See also Kuhn (1952) and Thackray (1970b).

The mechanistic view, whether Newtonian or Cartesian, appears to have been predominant in the early part of the eighteenth century. Thus, the two major French chemists were the Cartesians, Lemery and Geoffroy, and the major British chemists were the Newtonians, Keill, Freind, and Hales (Guerlac, 1951, p. 394; Schofield, 1970, p. 41). As might be expected, and as indicated by Guerlac, proponents of this view were not unreceptive to mathematics and quantification. However, their attempts did not amount to much. Boyle and others, for example, knew of the weight increase of metals in calcination and attributed it to the taking up of fire particles (Schofield, 1970, p. 41, 44). Hales, on the other hand, ignored this question and concentrated on measuring volumes of gases (Multhauf, 1962, p. 214). Both Guerlac and Multhauf point out that none of these chemists made essential use of quantification in any of their arguments: this development had to wait until the work of Black (Guerlac, 1961a) or Lavoisier (Multhauf, 1962).[2]

Another aspect of the corpuscular view, one potentially quantifiable, was based on the attractive forces between the ultimate particles. This was the notion of "affinities." Its first concrete manifestation was the 1718 table of Geoffroy, who was a Cartesian and used the French *"rapport"* in order to avoid assumptions about Newtonian attraction (Boas, 1959, p. 507). Duncan (1964, pp. 181-183) notes that it was the clear intent of this and all other early tables of affinities to *avoid any* theories and to emphasize the experimental nature of science.

The most influential chemist of the early eighteenth century was Hermann Boerhaave, who was important both for his writings and for his teaching.[3] His theories, which incorporated corpuscular ideas, were widely followed for many years (Rappaport, 1961, pp. 91-92).

However, attempts at quantification and explanation of chemical phenomena in these corpuscular terms failed, and by midcentury the climate of opinion had changed: mechanistic, Boerhaavian views lost their influence and those of the materialist Stahlians gained acceptance.

THE PHLOGISTIC REVOLUTION

The first paradigmatic theory of combustion and calcination — that is, the first to clearly associate these two phenomena and guide research — was proposed by Johann Joachim Becher (1635-1682) and developed and popularized by Georg Ernst Stahl (1660-1734). Stahl (1723) coined the term *phlogiston,* the matter

[2]The fact that there is disagreement about this basic question (was Black a quantitative worker or not?) indicates that there may well be problems in measuring this variable. See Appendix A.

[3]The influence of Boerhaave was so great that it made Leyden, where he practiced, the leading scientific institution in Europe in the early eighteenth century (Levere, 1970). As we shall see in Chapter 3, it was particularly important for the training of British physicians and chemists.

and principle of fire, contained in all metals and combustible bodies and given up in burning or calcination. The more phlogiston a body contains, the less residue there is when it has burned, so that charcoal or oils are viewed as almost pure phlogiston. Furthermore, the reduction of calces to metals when heated in the presence of charcoal is easily explained:

$$\text{calx} + \underset{\text{(charcoal)}}{\text{phlogiston}} = \text{metal}$$

The dating of the spread of Stahl's theory has undergone several changes,[4] but the present view is that it did not make much of an impact on French or British chemistry until close to midcentury. It appears that the path of its influence ran from Germany to France to Britain. Guerlac (especially 1959b) and his student Rappaport (1961), have made the case for France, and Schofield (1970) has done so for Britain.

Guerlac (1956) states:

> We should not forget that mineralogy and metallurgy had contributed greatly to the intensified interest in chemistry in France during Lavoisier's boyhood. Largely under the influence of German and Scandinavian writers, chemistry and mineralogy had come to be treated in close connection with each other; in fact this new emphasis had much to do with the general acceptance of the phlogiston theory during these same years. [p. 215]

As part of his analysis of the social and industrial influences on French chemistry during the 1750s and 1760s, Guerlac (1959b) treats the spread of this theory in more detail. He attributes the growing interest in chemistry in France to the close relationship it was seen to have with such industries as mining, metallurgy, dyeing, and pottery-making, and he points out the importance of the Baron d'Holbach's translations of German and Swedish mineralogical and metallurgical works. For these applications the theory of Stahl was eminently suitable, since it explained, for example, why all metals were alike: they all contained phlogiston. These industries were particularly important to France at this time, for in comparison to her major adversary, Britain, she was technologically backward.

While Guerlac emphasizes the social factors behind the spread of Stahlianism in France, Rappaport (1960, 1961) underscores the influence of Rouelle, who not only introduced Stahl's ideas into his popular courses of chemistry but also modified them to conform to his own background:

> While it is perhaps an exaggeration to say that Rouelle introduced a knowledge of Stahlian chemistry into France, he was certainly responsible for the popularization and eventual acceptance of a modified Stahlianism in the course of the century. [1961, p. 73]

---

[4]Compare, for example, Partington (1937), Kuhn (1952), and Guerlac (1959b).

She is careful to point out that while Stahl's work was known before Rouelle (e.g., see Senac, 1723), his system or theory was not:

> In the writings of Rouelle's immediate predecessors, chemists like Louis Lemery and P.-J. Malouin, Stahl is cited solely for specific points of method or for important experiments, but not for theory. [p. 93]

The major modifications Rouelle brought to Stahl's system were the incorporation of a French tradition (stemming from Geoffroy) of using affinities, the consideration of the role of air in chemical combination (due to Hales: Rappaport, 1961, p. 79), the examination of the (only apparent) weight gain during calcination of metals (Rappaport, 1961, p. 77), and a different definition of phlogiston:

> Stahl's phlogiston, a fatty earth, becomes associated in Rouelle's system with *fire* rather than with earth... The properties of phlogiston, as enumerated by Stahl, remain essentially unchanged in Rouelle's theory; phlogiston is still the principle of odors and colors, the principle released in combustions, and an element which may be set in motion by the action of the instrument fire. [p. 86]

It is this theory which replaced the previously dominant views of Boerhaave and his followers.

The revised version of Stahlianism produced by Rouelle achieved some theoretical economy over the "two-fluid" theory of Boerhaave, but Rappaport (1961) feels that its popularity was primarily a by-product of Rouelle's work, "both as a teacher and advocate of the theory and as a modifier and adapter of Stahl's ideas" (p. 92). After tracing the acceptance of Rouelle's theory through the works of Macquer, Rappaport concludes: "By 1766 ... Rouelle's modified Stahlian doctrines, also taught by his pupils Macquer, Venel, and Sage, were being widely promulgated and had become the dominant chemical system of the period" (p. 93). Thus, by shortly after midcentury, phlogiston had become the leading chemical theory of France.

Schofield argues that there was "creeping Stahlianism" in Britain in the second third of the century. He mentions Boerhaave's work, through the 1741 translation of his *Elementa Chemiae* by Shaw,[5] as the first step in this direction,[6] and goes on to point out that Shaw was also probably influenced by his first English translation of Stahl (1730). Through these influences, Stahl's ideas were introduced into British chemistry. As in France, the phlogiston theory was pre-

---

[5]This translation (Boerhaave, 1741) was disguised as "a second edition of the *New Method of Chemistry* with critical notes" (Schofield, 1970, p. 147).

[6]It is apparent that Schofield's interpretation of the work of Boerhaave differs from that of other historians, but the distinctions need not concern us here (see Schofield, 1970, pp. 146-156).

ceded by other ideas of Stahl:

> A survey of *Philosophical Transactions* articles on chemistry between 1731 and 1789 shows . . . that phlogiston becomes a significant part of British chemical explanation only gradually and, to any marked degree, after mid-century . . . Well before phlogiston seriously enters the scheme of chemical explanation, other parts of Stahl's chemical philosophy had already made their impact. [Schofield, 1970, p. 213]

While this quote indicates that phlogistic thinking entered Britain gradually, other evidence presented by Schofield shows otherwise. He notes (1970, p. 213) that the first "serious use" of phlogiston was by Huxham in 1754, "after which phlogistic explanations are fairly common." This suggests that there was a fairly rapid revolution in chemical thinking in Britain, as opposed to a more evolutionary process.[7]

Moreover, this revolution appears to have evolved from the one in France. Schofield notes the clear association of Stahl with French authors:

> When the transition began, about 1740, the standard authorities cited in British works on . . . chemistry were John Keill, Freind, and Boerhaave. The progress of change, by 1760, is indicated by the nearly unanimous substitution, for these authorities of . . . Stahl, Geoffroy, and Macquer in chemistry. [p. 191]

While phlogiston was certainly a major part of the paradigm which guided research after midcentury, and reflected the underlying materialist view of the structure of matter, including, in Rouelle at least, some of its quantitative aspects, it was not the only element of that paradigm. The other element, already seen in connection with Rouelle, was affinity theory, important in both France and Britain. Guerlac (1961a) points out that two of the earliest chemists to use affinities as the basis of their theories were Rouelle's student, Macquer, and Black's teacher, Cullen (see also Wightman, 1955, 1957):

> Though Macquer (in his *Elémens de Chymie Théorique,* 1749) was the first chemical pedagogue to build his basic theory on the doctrine of affinities, Cullen was the earliest to make 'elective attractions,' in the Newtonian sense, the fundamental doctrine of his teachings. [p. 208]

Affinity theory was not only an important explanatory tool, but also an element in attempts at quantification in the eighteenth century (Guerlac, 1961a, p. 208), reflecting its Newtonian heritage and the quantification of gravitational attraction.

LAVOISIER'S EARLY CAREER: ANOMALIES FOR PHLOGISTON

It is worth looking briefly at Lavoisier's early career and the explanations offered by historians for his ideas in order to grasp the intellectual and social

---

[7]We make this point because it shows that the rise of the phlogiston theory may also fit our model of science, which suggests that this process might be an appropriate test of our theory.

milieu in which Lavoisier began his work. It is well known that Lavoisier combined his interest in the phenomena of gases with a careful and essential use of the balance, striving to measure the weights of both the reactants and products of his chemical experiments, including the weights of the gases involved. However, the origins of his ideas and methods are not entirely clear. The situation was complex, and the records are incomplete, making reconstruction hazardous. Therefore, I will focus only on those elements of Lavoisier's theory that I think are especially important: the role of air in chemical combination, the structure of mineral acids, the states of matter, and the importance of weighing. (Compare Crosland, 1963, 1973; Gough, 1969; Guerlac, 1959a; Langer, 1972; Morris, 1972; Multhauf, 1962; Siegfried, 1972; Siegfried & Dobbs, 1968).

We know that Lavoisier began his career as a geologist, studying under Guettard (Meldrum, 1933, 1934). How did this prepare him for becoming a revolutionary chemist? Guerlac (1956) has reconstructed Lavoisier's education, emphasizing that Guettard was responsible for Lavoisier's apprenticeship under Rouelle, the instructor of perhaps the only course in chemistry that stressed geological concerns. We have seen that German mineralogical and metallurgical work had a strong influence on French chemistry and, especially, on Rouelle during the 1740s and 1750s. It is likely, then, that Lavoisier acquired his interests in German chemistry and mineralogy from Rouelle, interests which proved crucial to the development of his ideas (Gough, 1968; Guerlac, 1959a; Siegfried, 1972). Furthermore, Rouelle's view of chemistry not only included phlogiston but also, as we have seen (p. 25), the consideration of the role of air and weight gain during the calcination of metals.

Gough (1968) has discovered two manuscripts, written by Lavoisier in 1766, dealing with chemical theory. Based on a memoir by Eller in the *Mémoires* of the Berlin Académie for 1746 (published in French), they express Lavoisier's first interest in vapors or airs. Gough notes (1968, p. 57) that it was through Eller's memoir that Lavoisier became acquainted with the *Vegetable Staticks* of Hales (1727), the work Guerlac has shown (1951) to have been an important influence on Lavoisier in particular and on French chemistry in general. Gough further shows (1968, p. 55) that it was through the work of the German, Meyer, that Lavoisier became acquanted with the work of Joseph Black.[8]

Although the exact chronology of events is not clear — and is not essential to this study — several influences coalesced by early 1772 and precipitated Lavoisier's remarkable speculations in that all important year. Kohler (1972) and Crosland (1973) have written of Lavoisier's early, general interest in the composition of acids, "in keeping with the strong French interest in mineral resources and his own previous work on minerals and mineral waters" (Kohler, 1972, p. 350). Guerlac (1959a) and his followers (Gough, 1969; Morris, 1969,

---

[8]These findings not only exhibit the close relationship of German and French chemistry noted by Guerlac (to be further discussed in Chapter 6) but also at least the skeleton of an international chemical community.

1972; Siegfried, 1972) focused on Lavoisier's interest in airs. Not only was Lavoisier aware of the work done by Black and Hales on gases, but he was also interested in the role air plays in the effervescence observed when a metallic calx is reduced and converted into metal. This became familiar to him through the reading of Cramer's *Elementa artis docimasticae,* "a standard work on assaying, . . . translated into French in 1755" (Guerlac, 1959a, p. 119). Unpublished manuscripts uncovered by Fric (1959) bring out the importance of Lavoisier's interest in heat (Morris, 1969, 1972) and change of state (Siegfried, 1972). Lavoisier's manuscript of August, 1772, concludes with his "théorie singulière," the idea that air is not a simple substance but a *compound* of some particular fluid and the matter of fire − a radical break with the traditional four-element (earth, air, fire, water) theory, which was also part of Rouelle's phlogiston theory (Rappaport, 1961, p. 75). Another Lavoisier manuscript indicates that he was also aware of the work done by Cullen on evaporation and cooling, which demonstrated that gases are produced by heat, not the presence of air (Gough, 1969, p. 269).

The final significant influence on Lavoisier's interest in gas chemistry was the work of Priestley in the British pneumo-chemical tradition (Langer, 1972, p. 27), brought to France in 1772 through the efforts of Rozier and his *Introduction aux Observations sur la Physique.* To fully trace this influence requires some background material. It is generally accepted that the origin of interest in pneumatic chemistry was the work of Joseph Black on magnesia alba, which is tied to the earlier work of Hales. Cavendish, whom Guerlac calls "Black's first true disciple," continued the work on fixed air and other airs and published his findings in three papers in the *Philosophical Transactions* for 1766. Shortly thereafter, a "wave of enthusiasm" (Perrin, 1969, p. 141) for the "chemistry of airs" and the new discoveries and problems it uncovered[9] swept Europe and helped to swell the ranks of the chemists, leading to an international community:

> Men like Torbern Bergman and Carl Scheele in Sweden, Joseph Priestly in England, Antoine Lavoisier in France, and Felice Fontana in Italy were soon exchanging their ideas and experimental results among each other both publicly and privately. [Langer, 1972, pp. 27-28]

The growth of pneumatic chemistry was accompanied by the establishment of a major scientific journal in France, *Observations sur la Physique,* which became an international forum for this and other elements of the chemical com-

---

[9]The discoveries and problems uncovered by pneumatic chemistry created both interest and puzzles for the phlogiston paradigm, puzzles which could presumably be solved:

> These anomalies [reduction of mercury calx by heat alone and weight gain in calcination] posed no serious problem until an ardent Stahlian chemist [Morveau], attempting to tie up loose ends in the theory, summed up the evidence and clearly stated the problems, resolving both anomalies to his own satisfaction by inventive extensions of the phlogiston doctrine. [Perrin, 1969, p. 141]

munity. Priestley's work appeared there in 1772, and it reprinted Black's classic papers on magnesia alba in 1773 (with negative comments by the editor) as part of a debate between the proponents of Meyer's *acidum pinque* and those of Black's "fixed air" as a universal acid.[10] As we will see in Chapter 3, this journal devoted a larger portion of its pages to chemistry than either of the other two major French and British journals (the *Mémoires* of the Paris Académie and the *Philosophical Transactions* of the Royal Society of London). Furthermore, in contrast to those two journals, it had a truly international flavor, translating not only the articles of Priestley and Black of Britain, but also those of Meyer, Marggraf, and Achard of Germany, Bergman and Scheele of Scandinavia, and Fontana of Italy, among others. There is no doubt that this periodical, published at monthly intervals, was both a consequent and cause of the "wave of enthusiasm."

Thus, we have accounted for Lavoisier's interest in combustion and calcination, airs, and acids — the substances and processes of his interests. The other major element of the chemical revolution was Lavoisier's *method* — the emphasis on quantification through the use of the balance. It has long been thought that this emphasis came from Lavoisier's familiarity with Hales and Black, but some authors have suggested that Black was not the source (Gough, 1968; Multhauf, 1962), and Multhauf (1962) attributes it to a continental tradition:

> Whereas the use of the balance by the British chemists was occasional and incidental, it seems to have been used by the continental chemists with dogged persistence. . . As a product of the continental tradition, Lavoisier's use of the balance in his first published research was natural enough. [p. 217] [11]

Other evidence supports this latter suggestion: the work of Guyton de Morveau of Dijon, which demonstrated conclusively that all metals gained weight when calcined (reported to the Paris Académie in February, 1772, with Lavoisier present), the quantitative approach of Bayen and others in the *Observations sur la Physique* (Neave, 1951b, p. 144), as well as the examination of weight by Lavoisier's teacher, Rouelle (see p. 25 above). In a sense, then, Lavoisier fused the continental tradition of the use of the balance to the British tradition of pneumo-chemistry.[12]

---

[10]See Crosland (1973, pp. 309-310) for a discussion of this debate and its significance for Lavoisier's theory of acidity.

[11]According to Multhauf (1962, p. 217), the reason that Lavoisier was successful in quantitative interpretation, in contrast to the other continental chemists, was that he concentrated on *familiar* materials.

[12]The fact that use of the balance was part of the continental tradition may help explain why Lavoisier's new theory, based on such quantification, was resisted so long by British chemists (see Chapter 4 and 5): quantitative anomalies based on weight and quantitative arguments based on these anomalies had little effect.

Thus, although questions and controversy remain concerning the sources for and chronology of Lavoisier's early ideas and methods, this is probably not a mutually exclusive situation. It is more likely that all of these influences were in the continental milieu and available to Lavoisier. In any case, we do know that Lavoisier performed experiments on the burning of phosphorus[13] and sulphur in September and October of 1772 and showed that they gained weight. By the end of October he observed further that the reduction of lead calx with charcoal gave off an abundance of air. The result was the famous sealed note of November 1, 1772 (opened May 5, 1773):

> Sulphur, in burning, far from losing weight, on the contrary gains it . . . it is the same with phosphorus; this increase of weight arises from a prodigious quantity of air that is fixed during the combustion and combines with the vapors . . . what is observed in the combustion of sulphur and phosphorus may well take place in the case of all substances that gain weight by combustion and calcination; and I am persuaded that the increase in weight of metallic calces is due to the same cause. [quoted from McKie, 1952, pp. 72-73]

In his laboratory notebook for February 20, 1773, Lavoisier mapped out a series of experiments which he felt were "destined to bring about a revolution in physics and in chemistry." In two draft memoirs of Spring, 1773 (published by Fric in 1959) he developed his ideas on heat and change of state: (1) all bodies can exist in the three states depending on the amount

> . . . of fire with which a substance combines; and (2) *that the heat and light appearing during combustion derives largely, if not exclusively, from the release of the matter of fire combined with air,* not from an inflammable principle in the combustible substance . . . This second idea . . . was to provide Lavoisier with an important argument in his later open assault upon the phlogiston theory. [Guerlac, 1969, p. 382]

These speculations had arisen out of a phlogistic approach:

> On y voit qu'il [Lavoisier] est encore [in 1773] phlogistien, mais on sent aussi, qu'il commence à avoir des doutes sur la théorie de Stahl et sur les applications qui en ont été faites, et à entrevoir d'autres possibilités d'explication. [Fric, 1959, p. 137]

Siegfried (1972), in the most extensive article yet written on these early years, quotes Lavoisier's clear perception of anomaly and feeling of crisis (from the "Third Fric Memoir," assigned a date of April 21, 1773 (Fric, 1959, p. 166):

> My experiments are not complete enough yet to dare enter the lists against this famous chemist [Stahl]. Yet I believe I have said enough that one can sense that the present theory of the chemists is defective in many points, and that it is probable that the phenomena of fixed air when more thoroughly studied will lead this science to a time of almost complete revolution. [p. 77]

Although Lavoisier had perceived anomalies in the phlogiston theory which

---

[13]Mitouard's work on phosphorus was another influence of 1772 (Guerlac, 1959a, p. 135).

created a "crisis," at least for him, some time passed before he was able to launch a frontal attack on the phlogiston paradigm and foment a true scientific crisis.

DEVELOPMENT OF OXYGEN THEORY:
CRISIS AND PARADIGM DISPUTE

Lavoisier's first published results of his speculations on air and the problem of weight gain in calcination were his 1744 *Opuscules Physiques et Chymiques*. It took several years, however, for these doubts and speculations to converge into a new theory, and the first formal description of the oxygen theory came in papers presented to the Paris Académie des Sciences in 1777, in particular, the "Mémoire sur la combustion en général." His experiments, as well as those of Priestley and others, had made it clear that metals gain weight when calcined, that calces give up air when heated, and that the weight of this air is equal to the weight gained by the metal and lost by the calx. The same held true for other forms of combustion. Although Lavoisier had established in his own mind that the principle of combustion was the "eminently respirable" part of the air, not phlogiston, he was still hesitant about rejecting phlogiston theory completely.

During the same period Lavoisier and others, notably Berthollet, were doing experiments on the composition of acids, particularly the mineral acids[14] (muriatic, vitriolic, and nitric). According to Stahl's theory, as noted above, all acids were modifications of vitriolic (now called sulphuric) acid. However, experiments established that this was not the case, and in 1779 Lavoisier announced that each acid was formed by combination of a characteristic substance, or base, with the "eminently respirable" part of air, which he therefore named "oxygène" from the Greek, "I beget acid."

By 1783 Lavoisier had mounted an aggressive assault on the phlogiston paradigm. He stated in his "Reflections on Phlogiston," contained in the *Mémoires* of the Paris Académie for 1783 (published 1786), that phlogiston was an unnecessary assumption and that his theory was *simpler*. From this point on, he created critical problems for the proponents of the phlogiston paradigm. Among them were the problem of the gain in weight when a metal was calcined or some other substance burned: although some phlogistonists may have been able to ignore this, others felt it necessary to endow phlogiston with negative weight, the net result being an awkward diversity of explanations for this phenomenon. Another serious problem for the phlogistonists was that Lavoisier was able to isolate his principle, "oxygen," while phlogiston was either not isolable (Thackray, 1970b, p. 197) or was identified with various substances, such as "inflammable air" (now called hydrogen). Again, the anomalies led to contradictory explana-

---

[14]We have already seen that the French had a strong general interest in these acids and that Lavoisier had been concerned with them since early in his career.

tions by the phlogistonists. Finally, although less serious, his theory of acidity was superior to those offered by the phlogistonists (Crosland, 1973, p. 317).

However, he won few converts before 1785,[15] primarily because of some anomalies associated with his theory of acidity which derived from the "water problem," the question of the composition of water, an elementary substance according to traditional theory. The water problem was really two problems. First, when inflammable air (again, hydrogen) was burned, the product was not known. Furthermore, by Lavoisier's theory, the product should be an acid. The second problem involved the dissolution of a metal in an acid. Certain metals, when dissolved, give off inflammable air. When the solution is evaporated, a salt remains. This salt, when strongly heated, gives up the acid and yields a calx. When this calx is placed in acid, a salt remains and there is no inflammable air. The problem is the inflammable air: where does it come from?

The explanation is easy in terms of the phlogiston theory:

$$metal = calx + phlogiston \qquad salt = calx + acid$$

$$\underbrace{(calx + phlogiston)}_{metal} + acid = \underbrace{(calx + acid)}_{salt} + \underset{inflammable\ air}{phlogiston}$$

However, the new theory could not account for it: the solution to these problems had to await the discovery of the composition of water.

This discovery was the source of a major priority fight, the resolution of which has continued to the present day (see, for example, Perrin, 1973; Schofield, 1963a). The two principle contestants were Cavendish and Lavoisier. In 1783 Priestley noticed that a "dew" resulted when inflammable air (hydrogen) was burned in a globe. Cavendish found that the dew was "condensed" from inflammable air and one-fifth of common air and was, in fact, plain water. Cavendish did not publish quickly, which led to controversy over priorities.[16] The situation was complicated by the fact that when the inflammable air was burned in oxygen (then known as dephlogisticated air) instead of common air, the result was nitric acid (from the nitrogen left in the oxygen and the extreme heat of the flame). Cavendish "turned aside to investigate this and incidentally discovered the composition of nitric acid" (McKie, 1952, p. 117).[17] By the time his results were read to the Royal Society on January 15, 1784, they were rather well known. Not only had Priestley's results been reported to the Académie in

---

[15]Perhaps only Bucquet, Laplace, Meusnier — who worked with him — and Bayen. Some authors also suggest Black.

[16]The concern for priorities, which was extreme in Lavoisier (see Guerlac, 1961b, and the end of the "sealed note" of page 30), suggests that there was a desire for recognition among scientists at this time, supporting the general theory developed.

[17]Obviously, he did not discover the composition of nitric acid as we know it, for it contains oxygen, a substance in which Cavendish did not believe (cf. Partington, 1953).

May, 1783, but Blagden, president of the Royal Society, had communicated Cavendish's results to Lavoisier in June, 1783. Whatever the important factor (see Perrin, 1973, p. 427), Lavoisier had long been interested in this crucial reaction. Realizing the pertinence of these results to his theory, he immediately repeated the experiments together with the mathematician, Laplace, and reported the results to the Paris Académie on June 25. This he followed with some formal "claim-staking": on November 12, Lavoisier read a memoir on these experiments at a public meeting of the Académie which he quickly published in the December issue of *Observations sur la Physique*.[18]

In all versions Lavoisier gave little credit to Cavendish, whose experiments had been done first. However, the *explanations* were radically different from those of Cavendish, and this was clearly Lavoisier's triumph. The various phlogiston chemists explained the results in different ways, another sign of crisis in the paradigm, most of them trying to preserve the elementary nature of water. Lavoisier, however, saw water as a compound of oxygen and inflammable air, an interpretation which not only solved the problem of the result of burning inflammable air,[19] but also the problem of why the dissolution of metals in acids produces inflammable air:

$$metal + acid + \underbrace{(inflammable\ air + oxygen)}_{water}$$

$$= \underbrace{(metal + oxygen)}_{calx} + acid + inflammable\ air$$

$$= \underbrace{(calx + acid)}_{salt} + inflammable\ air$$

The composition of water was further demonstrated by decomposing it. Water was placed together with iron filings, which rusted, giving off inflammable air. The weight of the inflammable air plus the weight gain of the rusted filings was shown to be equal to the weight of the water consumed. Together with Meusnier Lavoisier carried out the "crucial" experiment in February, 1785, in

---

[18]The fate of this memoir illustrates a process which caused some difficulty, discussed in Appendix A, in dating some of the work done:

> The memoir was printed almost at once in the December issue of Rozier's journal *[Observations sur la Physique]*. Later it appeared in a more extended form in the annual volume of the *Memoirs* of the Academy for the year 1781, not 1783 as might have been expected, since the volume for 1781, with characteristic delay, did not appear until 1784. [McKie, 1952, p. 119]

[19]This explanation, however, does not solve the problem of why the product, water, is not an acid. This fact was conveniently ignored, just as the problem of weight was by many phlogistonists; no paradigm solves all questions that it raises.

which a large-scale synthesis of water was witnessed and controlled by a commission from the Académie.[20] The demonstration had the desired effect, for shortly thereafter the eminent and influential chemist Claude Louis Berthollet announced his conversion to the new theory.[21]

At this point the development of affinity theory and oxygen theory intersect. Berthollet and Morveau, two of the first converts to the new paradigm, were also two of the major developers of affinity theory, along with the Irishman, Kirwan (Guerlac, 1961a, p. 208; Wolf, 1952, pp. 378-379). Thus, the use of affinities and affinity theory was an integral part of both paradigms under consideration (see also Duncan, 1967).

CONVERSIONS: ACCEPTANCE OF THE NEW THEORY

Following the events of 1785 other important French chemists converted swiftly, including Fourcroy in 1786 and Morveau, when he came to Paris to discuss the new nomenclature in 1787 (Duveen & Klickstein, 1956). Conversions in other countries came more slowly, and accounts of resistance and reluctance to change are common.[22]

However, there is one phenomenon which has escaped the notice of historians and which seems to provide strong evidence for the "gestalt switch" nature of conversion: the importance of personal contact. Time after time we see that one chemist or another switched paradigms after personal discussion with a member of the opposite camp. While this does not seem to be as characteristic of the minor or younger scientists who are probably more receptive to new ideas (and about whom we get little information from historians), it is very important for major figures. The first converts were those who worked with Lavoisier; his collaborators − Bucquet before 1780 (according to McDonald, 1966, p. 74) and Meusnier and Laplace before 1785 − and Berthollet and Fourcroy, who worked with him on commissions shortly thereafter. Meusnier and Laplace also did work peripheral to the field of chemistry.

---

[20]Schofield (1964, pp. 289-290) denies that this was a crucial experiment. If what he means by that term is the deductive implication of the rejection of one theory and the acceptance of the other, obviously it was not: no experiment can be, as shown by Duhem many years ago, and I have adopted that theory here. But it clearly did function as crucial in the sense that it ultimately convinced at least one very important chemist (Berthollet) and probably others as well (Daumas & Duveen, 1959; Duveen & Klickstein, 1954a; Toulmin, 1957).

[21]For a textual study focusing on the actual date of Berthollet's conversion, see Partington (1959). Questions arise about the date because of the Paris Académie's practice of publishing its *Mémoires* some years in arrears.

[22]See Hufbauer (1971, pp. 118-180), for the first extensive treatment of the German case.

Morveau, as noted above, converted when he came to Paris for some months to discuss the new nomenclature and to witness experiments in Lavoisier's laboratory.

> After visiting Dijon [Thomas] Beddoes spent several weeks in Paris, and while under the influence of the French chemists he almost accepted the antiphlogistic theory, but he relapsed temporarily in 1789, and it was not until 1790 that he became completely converted. [Smeaton, 1967, p. 118]

Hufbauer (1971, pp. 172-173, 176) gives several examples of phlogistonists who converted after witnessing experiments performed by antiphlogistonists, although their phlogistonist teachers, who were not present, could not produce the same results.[23] Not all contact led to conversion to oxygen, however:

> Through his personal discussions with Priestley, Fontana, and Crawford, Kirwan apparently became convinced of the merits of their views and lent his active support. [Langer, 1972, p. 131]

> With the expressed purpose of winning over his colleague, Priestley traveled to London in March, 1783, and performed the experiment in Kirwan's laboratory; this time Kirwan was, for the moment, swayed in his opinion. [Langer, 1972, p. 170]

These examples indicate that it is easier to "see" what is going on if one can see someone else demonstrate that point of view and, perhaps, participate in that demonstration. Whatever the psychology of personal contact, it is clearly significant and a previously unrecognized phenomenon.

In addition to the evidence of "crucial" experiments, personal contact, and authority, there were other factors influencing conversion or resistance: nationalism (see Hufbauer on German resistance, 1971, pp. 127-131), metaphysical beliefs (see Schofield's (1970, pp. 260-262) explanation for why the materialist, Black, converted so easily while the mechanists, Priestley and Cavendish, never did), lack of convincing evidence (Schofield, 1964, p. 290, on Priestley), and the failure of either paradigm to answer all questions raised about it ("oxygen did not explain the mission of fire in combustion" (Schofield, 1964, p. 289); nor did oxygen account for the similarity among metals). Clearly, factors other than deductive logic were crucial in decisions to accept the new paradigm. To persist in viewing confirmation in science as a purely logical operation, as many philosophers do (e.g., Hempel, 1964; Scheffler, 1967), would be to place oneself

---

[23]Kuhn (1961) makes a similar point, talking about measurement:

> In scientific practice, as seen through the journal literature, the scientist often seems rather to be struggling with facts, trying to force them into conformity with a theory he does not doubt. . . Often scientists cannot get numbers that compare well with theory until they know what numbers they should be making nature yield. [p. 171]

In this case, each side knows what facts its theory says that nature should produce and finds that she does produce them.

in the ironic position of claiming that a major scientific advance was based on unscientific reasoning. It seems more accurate to conclude that nonlogical elements are inherent in the scientific method.

Three other major influences in favor of the new theory were: (1) the new nomenclature, (2) Lavoisier's *Traité* (1789), and (3) the establishment of a journal devoted to the new chemistry.

The new nomenclature, initiated by Morveau (who was influenced in turn by Linnaeus through Bergman (Smeaton, 1954a)), represented the combined efforts of the four outstanding chemists of France: Lavoisier, Morveau, Berthollet, and Fourcroy, as well as two younger men, Adet and Hassenfratz; it culminated in the *Méthode de Nomenclature Chimique* (1787). For the first time in the history of chemistry, the name given to a substance designated its chemical composition. Since the composition of a substance is theory-dependent, the nomenclature was a reflection of the theory, and its acceptance implied acceptance of the theory. This Lavoisier made clear in his presentation to the Académie on April 18, 1787:

> A well-composed language adapted to the natural and successive order of ideas will bring in its train a necessary and immediate revolution in the method of teaching and will not allow teachers of chemistry to deviate from the course of Nature; either they must reject the nomenclature or they must irresistibly follow the course marked out by it. [Quoted from McKie, 1952, p. 190]

Since it was the only systematic nomenclature for chemistry, the new system was a potent weapon in the struggle for adherents.

Two years later the new paradigm was presented systematically and fully in Lavoisier's *Traité Élémentaire de Chimie* (1789), complete with new nomenclature, apparatus, and experiments (exemplars). It did not suggest merely a new theory of combustion, but a new *chemistry;* not just a new theory of heat or acidity, but an altogether new view of chemical elements and their structure. Instead of the four traditional elements, there were now many of them, all defined in the materialist's terms: elements are those substances which have not yet been decomposed by the chemist's art, and the standard measurement for which is weight. Finally, the publishing of the *Traité* signified perhaps the first modern chemistry book: not a compendium of recipes or a dictionary of terms and subjects, but a system of chemistry laid out according to a central, unifying theory.

Although entitled a treatise, the book functioned essentially as a modern text: it exhibited the logic of the paradigm and showed how it was applied in concrete cases. The existence of a text is good evidence for the existence of a paradigm candidate (Kuhn, 1962, pp. 10, 155; 1961, pp. 163-167); it is the guide for puzzle solving in the new period of normal science. And in this case, it was clearly a force in consolidating the position of the new chemistry:

> By 1790 the new chemistry was on the offensive. Lavoisier's system was easier to work with, far easier to teach and to learn; through the *Méthode de Nomenclature* and the *Traité Élémetaire* it made a powerful appeal to uncommitted scientists, to

teachers, and particularly to students. After 1790, adherents flocked to the new chemistry. [Schofield, 1964, p. 287] [24]

Finally, the establishment in 1789 of the *Annales de Chimie* provided an outlet for proponents of the new theory, particularly younger men or foreigners, who could not publish in the *Mémoires* of the Paris Académie. This was particularly crucial, since de la Métherie, the editor of the major scientific journal at the time, the *Observations sur la Physique,* was a staunch opponent of the new theory. In addition to providing a place to publish, the *Annales* also fostered a sense of community.

The net effect was that the success of the oxygen paradigm was assured. In spite of diehards, especially in Britain and Germany, the younger generation, the generation of the future, was generally committed to the new theory, leading Lavoisier to write in 1791: "All young chemists adopt the theory and from that I conclude that the revolution in chemistry has come to pass" (quoted from McKie, 1952, p. 207).

Unfortunately, the French Revolution had also come to pass. Although its ultimate effect upon science seems to have been a substantial strengthening of French education, especially in science, its immediate effect was disastrous. Among the events associated with the Terror were the suppression of the Académie,[25] the resulting suspension of its *Mémoires,* the suspension of the *Annales de Chimie,* the suspension of or pressure put on other journals (the *Journal de Physique* did manage to continue publication, but at a slower rate), and the deaths of Lavoisier and Bailly (an eminent astronomer). These events may have affected the data and results during the later years of the Revolution (from 1793).

THEORY OF THE CHEMICAL REVOLUTION

Our theory assumes that the model of paradigm change is based on professional scientific communities; that is, groups of people trained as scientists by other scientists and engaged in science as a livelihood.[26] There is some question as to whether or not chemists in the eighteenth century fit this description. Most

[24]Both the kind of appeals mentioned and the kind of people being attracted fit our theory perfectly: this is no purely logical demonstration to the experts. On the influence of Lavoisier's *Traité,* see also Duveen & Klickstein (1954b).

[25]Although various *sociétés libres* did spring up, they were not very important during this period (Crosland, 1967, p. 175; see also Hahn, 1971, Chapter 9; Smeaton, 1957a, 1957b).

[26]The model may apply, however, even if these conditions are not met. In particular, scientists who are self-taught (from books, for example) or who are only "gentleman" scientists or parttime scientists may act sufficiently as "true" professionals for the model to be useful.

evidence seems to show, however, that for at least the last third of the century this is a reasonable assumption for France, although one perhaps less applicable to other countries,[27] including Britain.

While we will consider the two communities in more detail in Chapter 3, particularly their differences, it is necessary to say a few preliminary words here about their structure. The French chemical community (and, more generally, the French scientific community) had several elements of what we consider a profession: various training centers, including 22 universities, although not all of them had chemistry chairs (Costabel, 1964b; Dainville, 1964; Lacoarret & Ter-Menassian, 1964, p. 126); 16 medical schools on the eve of the Revolution (Huard, 1964, p. 172); and some technical schools, especially those of pharmacy (Bedel, 1964a; Birembaut, 1964; Huard, 1964; Taton, 1964, pp. 559-616), where chemistry was taught. While it is likely that the real training went on in *private* courses for promising students, it seems that not only did these university and technical school courses often introduce students to chemistry and introduce chemists to promising students, but they further served to provide salaries and laboratories, two prerequisites for *careers*. Moreover, some of these schools were associated with local scientific academies or societies (Costabel, 1964a, pp. 24-25; Smeaton, 1961), such as those at Montpellier and Toulouse, which not only provided for informal professional exchange, but also some formal exchange through the publication of memoirs.

We shall see, however, in Chapter 3 that the provincial schools and academies were not very important for French science; that is, they did not produce great numbers of chemists or articles, although they did make some important conttributions (Proust, 1968, esp. Chapter 5). Of overwhelming importance was Paris, a city unique in the world for the variety and richness of its scientific institutions. Not only was Paris blessed with several major schools[28] and societies,[29] but also the Jardin du Roi or Jardin des Plantes, where some of France's (and

---

[27]For example, Hufbauer (1971, Chapter 1) shows that there were few training grounds or paying positions for chemists in Germany until late in the century, when a few chairs were established in the universities and the Berlin Academy of Sciences paid a few of its members.

[28]Among these were the Université de Paris (Lacoarret & Ter-Menassian, 1964, pp. 126ff), the Collège Royal with a chair of chemistry and pharmacy (Huard, 1964, p. 184) which became chemistry alone in 1772 (Torlais, 1964, p. 267), and the Faculté de Médecine, where Augustin Roux was given the first chair of chemistry in 1771 (Huard, 1964, p. 180).

[29]Among these were the Académie Royale des Sciences, with a chemistry section dating from its reorganization of 1699, the Société des Pharmiciens, and the Société Royale de Médecine. During the 1780s and 1790s societies proliferated rapidly, as they did in the rest of France and other parts of Europe.

Europe's) most famous teachers of chemistry[30] taught free courses to the masses (including ladies and gentlemen of fashion).

Therefore, in France in general and in Paris in particular, a competent student of chemistry could pursue a chemical career with some assurance of employment. For those who achieved election to the Académie, there were further rewards: an eventual stipend as a "pensionnaire" and facilities for work. Furthermore, Academicians (and, to some extent, others) received support from the government in the form of positions as directors of various industrial enterprises (Guerlac, 1959b, pp. 94-96; Hahn, 1971, pp. 69-72). And, as we shall see in Chapter 3, there were readily available outlets for scientific production and recognition among the various scientific journals published in France, particularly the publications of the provincial and Paris academies, the *Observations sur la Physique,* and other, more specialized journals, such as the *Journal de Médecine.*

In Great Britain, conditions were quite different. In England proper the two universities, Oxford and Cambridge, were dead as far as chemistry was concerned: there were no paid chairs, so the teachers were either incompetent or nonprofessional, sometimes both, and unassociated with scientific societies, and since chemistry was not required, it attracted few students (Coleby, 1952, 1953, 1954; Gibbs & Smeaton, 1961; Hans, 1951; Levere, 1970). And in London, where the Royal Society was located, there was no university or other educational institution (aside from itinerant lecturers) where a chemist could find employment. Furthermore, the Royal Society itself was at a low ebb,[31] and its members were "amateurs" in the worst sense of the word, often merely gentlemen with no scientific qualifications whatsoever.

In Scotland the situation was somewhat better, for there were some schools, notably the medical schools at Glasgow and Edinburgh, where chemistry was taught (Hans, 1951).[32] Also at Edinburgh was the scientific society which eventually became the Royal Society of Edinburgh. So Edinburgh was the closest

---

[30]The Jardin du Roi was endowed with two chairs of chemistry, a professor and a demonstrator, from 1695 (Laissus, 1964, p. 312). Teachers there included G. F. Rouelle, H. M. Rouelle, and Fourcroy. Another influence was private museums or "cabinets" (Bedel, 1964b).

[31]Contemporary accounts often made mention of this fact, particularly the reviews in the German literature. Compare also Multhauf (1966, p. 305), Trengove (1965, p. 184) and Schofield (1966, p. 145). Hahn (1971, p. 137), curiously, seems to disagree — see p. 66 below.

[32]Not only did two of the great teachers of chemistry, Cullen and Black, practice at these schools, but their scientific tradition stemmed from close, long-standing ties with Leyden (Levere, 1970, esp. p. 46; Wightman, 1955, 1957).

approximation to Paris in Britain:

> In the late eighteenth century only Edinburgh among English and Scottish centers possessed a thriving university and an active scientific society. [Morrell, 1971, p. 166n]

Toward the end of the century, other academies were established in Britain, as they were in France, usually in the new industrial towns such as Manchester and Birmingham (Musson & Robinson, 1961; Robinson, 1953, 1955, p. 137). Still, these groups were not significant during the period under study, at least in terms of the number of chemists there who published.

There were also differences in the structure of formal organizations in France and Britain. French scientific societies were usually modeled after the Paris Académie (often referred to as the "academy" model of organization), that is, a hierarchical organization, with ranks and restricted membership. British scientific organizations, on the other hand, usually resembled the Royal Society of London (often called the "society" model of organization), in which all members were peers and to which election was relatively easy (McKie, 1948b). A further distinction was that the Paris Académie required members to live in Paris and attend meetings regularly[33]; the Royal Society of London did not restrict its members in any way.

While there were exchanges of remuneration between science on the one hand and technology and industry on the other in both France (Gillispie, 1957; Guerlac, 1956, 1959b; Hahn, 1971, p. 69) and Britain (Clow & Clow, 1952; Cochrane, 1957; Gibbs, 1952b; Schofield, 1959, 1963b; Thackray, 1970a; 1970b, pp. 107, 204), these links were different in the two countries. In Britain, especially, as we shall see in Chapter 3, they seem to have been a two-edged sword.

In sum, then, there was a fairly strong and well organized chemical community in France in the late eighteenth century, consisting of educational institutions, respected scientific societies, employment possibilities for chemists, and periodicals for scientific communication (and control). In Britain, however, the organization was decidedly weaker, comprising few educational institutions for either training or employment, a major scientific society in disrepute and others just beginning, although there were a few journals and some employment opportunities existed in industry.

---

[33]One of the consequences of this requirement was to keep certain nationally prominent men, such as Guyton de Morveau, from becoming members if they did not wish to come to Paris to live. Therefore, although the Paris Académie was a "national" institution in the sense that it dominated French science and was looked to for leadership by provincial academies (Proust, 1968, p. 9; Hahn, 1971, p. 89), it was not national in the sense of the Royal Society of London, which had as full members men from various parts of Britain as well as from foreign countries. The Paris Académie did have, however, special categories of foreign and non-Parisian members and correspondents as well as a special connection with the Académie at Montpellier (see Proust, 1968, p. 9, for further details on the connection).

In any case, I will assume that there was a reasonable semblance of a professional scientific community in France,[34] and enough of one in Britain (there were chemists, and there were journals) to merit investigation. In Chapter 3 I will treat this issue in more detail.

The following facts have now been established:

(F1)   A paradigm based on phlogiston was dominant in the 1760s.

(F2)   The Chemical Revolution involved a switch from a paradigm based on phlogiston to one based on oxygen in an attempt to explain the phenomena of combustion, calcination, and acidity.

(F3)   The Revolution also involved a change from qualitative (or "pseudo-quantitative," such as recipes) work to a basic reliance on the balance.

(F4)   Phlogiston chemists tended to avoid quantitative arguments.

(F5)   Affinities and/or the affinity theory were a significant part of both traditions and led to quantification.

(F6)   The new paradigm was due primarily to the work of a Frenchman, Lavoisier, and the Revolution began in and spread from Paris.

(F7)   The following sequence of events occurred: 1772-77, anomalies became apparent; 1778-84, a crisis was created and a new paradigm was proposed; 1785-89, crisis and paradigm debate became widespread, conversions began in France; 1790-95, the new paradigm became institutionalized in France (most British chemists had not switched by 1795 — see Chapter 4).

Combining these facts with the propositions put forth at the end of Chapter 1, we derive the following set of testable hypotheses about this scientific revolution:

(H1)   The proportion of journal articles using phlogiston will decline, somewhat erratically at first, beginning about 1772, and the proportion of antiphlogiston or oxygen articles will increase so that by 1795 there will be very few phlogiston articles published (from (F2) and (F7)).

(H2)   The proportion of theoretical articles will increase until some time between 1785 and 1789, then decrease; the proportion of theoretical work will be high among oxygen papers at first and decline as the oxygen paradigm becomes accepted (from (1), (F2), (F7)).

(H3)   The proportion of quantitative articles will increase, especially after 1777; the proportion of quantitative articles among papers using oxygen will be

---

[34]Multhauf (1962) provides support for this assumption:

The typical eighteenth century chemist on the continent turned to analysis . . . Broadly understood as the complete description of substances. . . [They worked on the] principle chemical materials of commerce. . . The usefulness of these analyses soon manifested itself to the monarchs who supported the continental societies, and we encounter individuals who in the last half of the century look very much like professional chemists. [p. 217]

very high at first, and the proportion of quantitative articles among phlogiston articles declines after about 1785 (from (3), (F2), (F7)).

(H4)   The more theoretical the articles are, the greater the proportion which will be quantitative after 1777 (from (5) and (F7)).

(H5)   There will be a larger proportion of oxygen articles among French journals than among British ones at any point in the Revolution (from (F6)).

(H6)   The proportion of papers using the oxygen paradigm will be higher among the more theoretical papers than among the less theoretical (from the discussion of relationship of anomalies to theory (pp. 15-16), (F2), and partly as a consequence of (1b)).

(H7)   The proportion of papers using the oxygen paradigm will be higher among quantitative papers than among qualitative (from the discussion of the relationship of anomalies and quantitative work (pp. 16-17), (F2), and partly as a consequence of (3b)).

(H8)   The number of authors will increase rapidly after 1777 (and may also increase rapidly after 1772 due to the great interest aroused by the new gas chemistry) (from (6) and (F7)).

(H9)   The productivity of the authors will increase as the Revolution progresses, especially after 1777 (from (7) and (F7)).

(H10) The number of neutral papers will be small (from (10)).

(H11) French articles will shift to the oxygen paradigm before the British (from (F6)).

(H12) The shift to the oxygen paradigm will occur earlier among the more theoretical papers (from (2) and (F2)).

(H13) The shift to the oxygen paradigm will occur earlier among quantitative papers than among qualitative ones (from (4) and (F2)).

(H14) Younger authors will shift to the oxygen paradigm earlier than older authors (from (8), (9), (F2)).

(H15) The younger an author is, the more likely it will be that he will use the oxygen paradigm once the choice is available (from (8), (9), (F2)).

(H16) There will be a positive association between theoretical papers and use of affinities (from the discussion of affinities (p. 26) and (F5)).

(H17) There will be a positive association between quantitative papers and use of affinities (from (F5)).

(H18) The proportion of citations to authors who supported the oxygen paradigm will increase, and the proportion of citations to authors who use phlogiston will decrease as the Revolution progresses, especially after 1785 (from (11), (F2), (F7)).

The following hypotheses are more conjectural than absolute, for they are based upon assumptions about the countries involved and do not depend only upon the theory I have developed:

(H19) The proportion of papers which are theoretical will be higher among French than among British articles, and this difference will increase as the Revolution progresses.

(H20) The proportion of papers which are quantitative will be higher among French than among British articles, and this difference will increase as the Revolution progresses.

3

The Chemical Community

Science, in general, and chemistry, in particular, became more established as a professional career during the latter part of the eighteenth century. The flowering of scientific societies had begun somewhat earlier during the Enlightenment (McKie, 1948b; Proust, 1968), and professional scientific journals now began to flourish (McKie, 1948a). A crucial development, which had started in France early in the century at the Paris Académie des Sciences (Hahn, 1971) and spread elsewhere in Europe (Guerlac, 1959; Hufbauer, 1971; Taton, 1964; Taylor, 1948), was the establishment of salaried positions for scientists in universities, academies, and government bureaus. Chemistry was a major beneficiary — we can point to such examples as Rouelle at the Jardin du Roi, Lavoisier at the Arsenal, and Berthollet at the national dyeworks. We shall see that Britain lagged in this respect (cf. Crosland, 1962).

In this chapter I will evaluate the hypotheses on productivity, (H8) and (H9), summarize the character of the journals, describe the productivity of authors, and account for differences in the social structure of the French and British scientific subcommunities. I will also depict the chemical communities of authors and their channels of communication. This will set the stage for hypothesis testing.

PRODUCTIVITY

Between 1760 and 1795 French chemical authors outnumbered their British counterparts by more than 3 to 1 (159 to 48, see Appendix A). Furthermore, British articles published on chemistry (134, or an average of 3.7 per year) amounted to only 19% of the French articles published (724, or an average of 20.1 per year). These figures are surprising: although it is known that French science, in general, and French chemistry, in particular, were superior to British, the magnitude of the difference is hardly hinted at in the literature. It is true that we have seen (p. 39) that the Royal Society was not what it had once been and that science was practically dead in Oxford and Cambridge, while the Paris Académie was preeminent in science and chemistry flourished at various educational institutions in France; yet we have also seen that Edinburgh was a leading scientific center, and that many notable chemical discoveries were made by British chemists. Thus, the first important finding of this study is that there was an enormous quantitative difference in periodical productivity between the French and British chemical communities, both in the number of authors publishing and even more so in the quantity of articles published. The distribution of articles over these time periods may provide some clues to the differences between the two communities and test the productivity hypotheses, (H8) and (H9). Tables 3.1 and 3.2 present the results for the basic and cited samples, respectively. Although the cited (recognized) group is our main interest, consideration of the basic sample will show the wider attraction of chemistry.

The data in Table 3.1 generally support the theory that there was a gradual increase in both the number of authors publishing (H8) and the number of articles published (H9) from 1760 to 1765 until 1793-95 in France and 1791-92 in Britain, although these increases were irregular.[1] There are some differences between the two countries, however, which merit discussion. These relate to the

---

[1] A few remarks are in order regarding the calculation of these rates. "Authors publishing per year" is an average of the number of authors publishing in each year; that is, an author is counted only once for each year he published. The sum of the authors counted in this way is divided by the number of years in the period. "Articles per year" is calculated by adding up all the articles published in a given period and dividing by the number of years in the period. "Articles per author per year" is then calculated by dividing the articles per year by the authors publishing per year, which gives the average number of articles published each year by authors who publish. This method can produce strange results: if an author published 5 articles, 1 in each of 5 years, then he would be counted as a publishing author 5 times for an average of 1 author/year over the 5 years. On the other hand, if he published all 5 articles in 1 year, he would be counted only once as a publishing author for an average of .2 authors/ year over the 5 years. Since he published 5 articles in 5 years, the articles/year would be 1 in each case, resulting in productivity (articles/author/year) of 1 in the first case and 5 in the second. The differences in pattern could be useful information. However, this means that we must be careful also not to divide an artificially low author/year rate into articles/year, thus producing an artificially high productivity rate. With this in mind, I rechecked the data and there was no evidence of such artifacts. In fact, Tables 3.1 and 3.2 both show that productivity is *high* when authors/year are also *high* (and articles/year as well). Another way to handle the problem would be to think in terms of a "pool" of talent and calculate "author-years" much as the demographer calculates "person-years" or "man-hours." This would include authors who are alive but who did not happen to publish in a given year. In fact, we did this for the Academicians discussed in relation to Table 3.13 and got results similar to the ones presented. However, it is very difficult to determine what the appropriate "pool" would be when some authors publish frequently, some only once, some as chemists, and some as doctors or pharmacists.

TABLE 3.1
Distribution of Authors and Articles over Time, Basic Sample

	1760 -65	1766 -71	1772 -77	1778 -80	1781 -84	1785 -86	1787 -88	1789 -90	1791 -92	1793 -95
				Authors publishing per year						
French	4.8	6.0	13.5	18.7	15.0	19.0	14.0	19.5	21.0	8.0
British	0.7	2.3	2.5	2.3	2.8	2.5	6.5	8.5	6.0	2.3
B/F[a]	(14)	(39)	(18)	(13)	(18)	(13)	(46)	(44)	(29)	(29)
				Articles per year						
French	5.8	6.7	18.3	30.3	26.0	42.5	28.5	38.0	46.5	11.0
British	0.7	2.7	3.0	2.3	3.5	4.5	9.5	13.0	6.5	2.7
B/F	(11)	(40)	(16)	(8)	(13)	(11)	(33)	(34)	(14)	(24)
				Articles per author per year						
French	1.2	1.1	1.4	1.6	1.7	2.2	2.0	1.9	2.2	1.4
British	1.0	1.2	1.2	1.0	1.3	1.8	1.5	1.5	1.1	1.2
B/F	(83)	(104)	(88)	(62)	(75)	(80)	(72)	(79)	(49)	(85)

[a]B/F = British rate as a percentage of French rate; ( ) = percentages. Articles per author per year are the average number of articles per publishing author per year (articles per year divided by authors publishing per year).

size of the numbers and the timing of changes (increases and decreases). After looking at the data for each country, we will compare the two.

The French data show good overall support for the two hypotheses. Except for the last period, when the events of the French Revolution held sway, there was a rise in authors publishing per year and articles published per year: from a low of 4.8 in 1760-65 the average number of authors publishing per year jumped in 1772-77 to 13.5 and continued erratically to a peak in 1791-92 of 21.0; and the number of articles published per year increased from a low of 5.8 in 1760-65 to a peak of 46.5 in 1791-92, again jumping sharply upward in 1772-77. In both cases there were secondary peaks in 1785-86, which we must attribute to the large number of conversions to the new theory: not only was this period the peak of crisis and debate, but new converts would doubtlessly be especially vigorous in demonstrating their new faith. Since the decline after 1792 can be attributed to the suppression of academies and journals during the Terror, our prediction of an increase in the number of authors and in publications accompanying a paradigm dispute (H8) appears to be supported by the figures, although the increase may also be due to the long-term secular growth rate of science (see p. 50).

Looking at *productivity* of the authors (articles per author per year), we find a similar pattern. After a slight decline in 1766-71, this rate increased smoothly to a peak in 1785-86 of 2.2, declined slightly (1.9) until 1791-92, and then returned to 2.2. A slight decline after the crisis stage has passed to the consolidation stage is not surprising, for it probably represents a short period of intellectual and emotional regirding before the exploitation of the new paradigm. Having demonstrated their faith, the new converts may temporarily relax. The evidence from this one case, of course, is statistically weak and merely suggestive, and the question bears further study.

In the British data we find similar patterns, with some minor differences in timing. Again there was an increase over time in the number of authors publishing per year (from 0.7 in 1760-65 to 2.3 in 1766-71 to a peak of 8.5 in 1789-90), in the number of articles per year (from 0.7 in 1760-65 to 2.7 in 1766-71 to a peak of 13.0 in 1789-90), and in productivity (from 1.0 in 1760-65 to a peak of 1.8 in 1785-86 declining to 1.2 in 1793-95). At first glance, the British data appear to support the theory more strongly than the French. The high productivity of authors in 1785-86 was probably a response to the challenge of the oxygen chemists. The sharp increase in 1787-90 was due in part to a crisis: the argument over the composition of water and nitric acid. However, the new paradigm was

TABLE 3.2
Distribution of Authors and Articles over Time, Cited Sample

	1760 -65	1766 -71	1772 -77	1778 -80	1781 -84	1785 -86	1787 -88	1789 -90	1791 -92	1793 -95
Authors publishing per year										
French	3.8	4.3	11.0	10.0	12.0	13.0	13.5	15.0	17.0	6.7
British	0.3	1.7	2.0	1.7	2.8	2.0	4.5	6.0	4.0	1.0
B/F[a]	(8)	(40)	(18)	(17)	(23)	(15)	(33)	(40)	(25)	(15)
Articles per year										
French	4.5	4.8	15.3	24.0	22.8	37.5	25.5	32.5	42.0	10.0
British	0.3	1.7	2.5	1.7	3.5	4.0	7.5	9.5	4.5	1.3
B/F	(7)	(35)	(16)	(7)	(15)	(11)	(28)	(29)	(11)	(13)
Articles per author per year										
French	1.2	1.1	1.4	2.4	1.9	2.9	1.9	2.2	2.5	1.5
British	1.0	1.0	1.3	1.0	1.3	2.0	1.7	1.6	1.1	1.3
B/F	(83)	(91)	(93)	(42)	(68)	(69)	(90)	(73)	(50)	(88)

[a]B/F = British rate as a percentage of French rate; ( ) = percentages. This table is based on 616 French and 101 British articles.

almost universally rejected in Britain at this time, so we cannot attribute any decline to a postconversion effect. On the contrary, it is likely that the decline in Britain was due partly to the effects of the French Revolution, as we shall see in the discussion of Table 3.2.

The superiority in numbers of French authors and articles previously observed existed in all time periods, although the large percentage increase in authors and articles began a few years earlier in Britain (1766-71) than in France (1772-77). Why relative British production of authors (14%) and articles (11%) was so comparatively low in 1760-65, we cannot say, but the subsequent rises and declines relative to the French are explicable in terms of the theory and known historical events. The rise in 1766-71 to 40% appears to be a direct consequence of the British superiority in "pneumatic" chemistry, and we attribute the subsequent decline to the development of pneumatic chemistry in France, to the establishment of the *Observations sur la Physique,* and to the attack on the phlogiston theory by Lavoisier and the consequent response by the French chemical community. These events culminated in the conversion of the major French chemists to the new theory in 1785-86. The relatively large fraction of British authors (45%) and articles (34%) in 1787-90 was due to the water controversy, as we shall see in the discussion of Table 3.2.

While these data do tend to support the theory, we should refrain from concluding too much, since we are dealing only with the basic sample, which includes many authors who were not chemists in the usual sense of the word (they published articles with some chemical content, often minimal) and about whom we have practically no information. Far more significant are the patterns of those men who were "recognized" (cited) by the community as having made contributions, and for the remainder of the chapter I will discuss the "cited" sample. (Appendix C gives the distributions for the basic sample.)

In this sample, too, French overwhelmingly dominate in terms of numbers. There were only 26% as many British authors (27) as French (102), and the British authors published only 16% as many articles (101) as French authors (616). On the average, a British author published only 52% as many articles (3.74) as a French author (6.04). This indicates that the great difference in the number of articles is due primarily to the greater number of French authors and secondarily to the greater productivity of French authors. Thus, we should look for the reasons why there were so many more French chemists than British, and why they were significantly more productive. We shall take up this question after we look at the distributions over time and their bearing on our theory.

As in the first sample, the distributions of French and British data over time generally support the theory: both the number of authors publishing per year (H8) and the number of articles published per year (H9) increased from 1760-65 through 1791-92 in France and through 1789-90 in Britain, then declined. Again, there were differences between the two countries' numbers and timing, reflecting their different scientific climates.

The growth in France began shortly after 1771, coinciding with the great wave of enthusiasm for the chemistry of "airs"[2] and the establishment of the *Observations sur la Physique,* and declined only after the suppression of the Académie and various scientific journals during the Terror in 1793. With only minor fluctuations, there was a steady growth in the number of chemists publishing per year, from 3.8 in 1760-65 to 11.0 in 1772-77 to a peak of 17.0 in 1791-92. The growth of articles was more spectacular, although somewhat erratic. From 4.5 in 1760-65 it rose sharply in 1772-77 to 15.3 and again in 1778-80 to 24.0, strongly supporting our prediction of a rise in productivity after 1777 (H8). The difference between the parallel rise of authors and articles in 1772-77 and their divergence in 1778-80 bears examination. The rise in authors and articles in 1772-77 was probably largely due to the attraction of pneumatic chemistry to potential scientists (see also the discussion of new authors below), while the divergence in 1778-80 suggests that those men who were already chemists were responding to the attacks by Lavoisier on their paradigm, although part of the rise in articles was due to Lavoisier himself. Once the excitement of the crisis spread, there was steady growth in the number of authors, yet it was not as marked as the rise in the number of articles, which peaked at 37.5 in 1785-86 when the paradigm dispute was at its height and the new theory began to win major converts. Subsequently, there was a mild decline in the number of articles per year, followed by an increase to a high point of 42.0 in 1791-92, when the French Revolution intervened. The establishment of the *Annales de Chimie* in 1789 undoubtedly contributed to the resurgence of the pattern of growth.

Variations in the *productivity* (articles per author per year) of French chemists are similar to those in the number of articles, except that the absolute peak of productivity occurred in 1785-86 rather than 1791-92. From 1.2 in 1760-65 there was a slight increase to 1.4 in 1772-77, and a big jump to 2.4 in 1778-80, predicted by (H9), reflecting Lavoisier's initial assault on phlogiston chemistry and the subsequent reaction of phlogiston chemists (as the data of Chapter 4 will show more clearly). The highest level, 2.9, coincided with the peak of debate and the "crucial experiment" demonstrating the composition of water. This was

---

[2]The great enthusiasm for pneumochemistry is not unrelated to our theory, for one could make the case that this constituted a minor revolution – in method alone, if nothing else. Since it was in many ways a new field, many of the elements of crisis were present. Specifically, there was a great opportunity for a young chemist to establish himself with little risk: there were no recognized experts to compete with, for Black did not publish from 1756 to 1794, Cavendish had not yet made his mark when British productivity jumped in 1766-71, and there were no French experts at all when the French took up the cause in the early 1770's (cf. p. 18).

also followed by a slight decline and then another increase to 2.5 in 1791-92. While the number of *cited* French authors publishing per year did not crest in 1785-86, as did the *total* number of French authors, both the number of articles and the number of articles per author did, a peak which may be attributed to the effects of crisis and the initial conversions of outstanding chemists.

While there was an increase in the number of authors and articles over time in France, as predicted by the theory (which also predicts much of the timing of the increase), there were factors other than the Chemical Revolution which could have affected these data, although most characteristics of the patterns over time still support our view. The distribution of articles and authors over time may also be connected to the great interest in pneumatic (gas) chemistry in the early 1770s (see footnote 2, p. 49), and the long-run secular growth rate of science, as well as the increasing professionalism of chemistry (Hahn, 1971, p. 276), independent of the Revolution itself. Probably, however, the timing of the variations in the rates is best explained by our hypotheses. One complicating factor, resulting in decreases after 1792, was the French Revolution. Its effects are still not entirely clear (Gillispie, 1959; Hill, 1959; Guerlac, 1959c; Williams, 1959), although it is obvious that one short-term result was the suppression of the Académie and other forms of scientific organization.

The data on productivity per author provide further support for the explanation/prediction of an increase. First, the large increase did not occur until 1778-80, an increase more likely due to effects of the growing crisis in the old paradigm than to the interest in pneumatic chemistry which had occurred a few years earlier.[3] Second, the zenith was in 1785-86; this is predictable from the discussion of conversion in Chapters 1 and 2, for these were the years of the first major conversions (Berthollet and Fourcroy).

The British data show similar patterns, although with differences in timing. The number of articles published grew from 0.3 in 1760-65 to 1.7 in 1766-71, 1766-71 and remained fairly constant until a sharp increase to 4.7 in 1787-88, after which it peaked at 6.0 in 1789-90 and declined rapidly to 1.0 in 1793-95. The number of articles published grew from 0.3 in 1760-65 to 1.7 in 1766-71, followed by fluctuating growth until it again climbed steeply to 7.5 in 1787-88, crested at 9.5 in 1789-90, and declined sharply to 1.3 in 1793-95. Productivity (articles per author per year) took a slightly different course. From 1.0 in 1760-65 it fluctuated until 1785-86, then rose sharply to 2.0 and remained relatively high (1.6) until 1790, when it declined along with the other rates. The jump in 1766-71 was a manifestation of the flowering of pneumochemistry, which was begun in Britain by Black and his followers. There followed a period of slow growth until another large increase in the late 1780s. If we look more closely at the individual articles involved in this burst of activity, we see that it was primarily due to the debate over the composition of water and nitric acid, two sub-

---

[3]The finer data of Chapter 4 provide more support for this line of reasoning.

stances which became cornerstones of the new paradigm,[4] and to the initial publication of the *Memoirs* . . . of the Manchester Literary and Philosophical Society.

Surprisingly, the British data exhibit a decline in production in 1791-92 (noted in the discussion of the basic sample), earlier than the French data. Although we do not have enough information to be sure, the decline was very likely due, in part, to the effects of the French Revolution in Britain. For example, the Birmingham riots of 1791 had caused Priestley, Britain's most prolific chemist and an outspoken supporter of the Revolution, to abandon his house, curtail his work, and, eventually, to depart for America, where he again took up the battle for phlogiston until his death in 1806. It is also likely that the (perhaps not coincidental) decline of the Lunar Society, which occurred at this time (Schofield, 1963b, pp. 190, 372-373), was a contributing factor. While we have not singled out this society for special treatment as we have the Paris Académie and the Royal Society of London, some notable British chemists were among its members, including Priestley, and several works of Schofield (1957a,b; 1963b; 1966) have demonstrated the Society's importance in British chemistry.

Returning now to French-British comparisons, it is obvious from Table 3.2 that the number of British chemists was only a fraction of the French at every time interval. There are, nonetheless, clear patterns in their relative strength (or weakness). Looking at the "B/F" lines of the table, we find the greatest representation of the British authors came at two points: first, in 1766-71, when they took the initiative in gas chemistry and contributed 40% of the manpower and 35% of the articles; and second, during the nitric acid-water dispute in 1787-1790, when they produced 40% (1789-90) of the manpower and 29% of the articles. The low points of British representation are also related to the processes we have been studying. Practically no chemical research was being done in Britain during 1760-65. Once the French chemical community responded to Lavoisier's questions and developments in gas chemistry, British representation plummetted, reaching a low of 7% of the French rate of article production in 1778-80 and 15% of French numbers of authors in 1785-86. After the "recovery-by-crisis" in the late 1780s, their relative production of articles dropped once again, to 11% in 1791-92. Productivity per author, while more respectable, also reached relative lows in 1778-80 (42%) and 1791-92 (50%). While the percentages were

[4]The demonstration of the composition of these two substances provided two *paradigms* in the sense of Kuhn's concrete puzzle-solutions or models (1970, p. 175). While the analysis and synthesis of water, according to principles of the new theory, led to swift victory for the paradigm in France, it appears to have merely ignited the crisis stage in Britain. Undoubtedly, some of the other fluctuations in the data are due either to disputes or to the publication of new journals appearing (or to journals not being published every year). For Britain these fluctuations can result in large differences because of their correspondingly small numbers, but among French articles these factors are probably overridden by larger effects.

higher and the timing only slightly different for relative productivity, we must consider further reasons for the remarkable weakness of British chemistry.

We initially suggest that the great quantitative gap between British and French productivity was caused by differences in the organization, training, and facilities in science mentioned at the end of Chapter 2.[5] Some of the divergence may also be due to differences in publication patterns between the two countries. It could be, for instance, that British chemists published overwhelmingly in books rather than in journals. It is well known that books and monographs were important forms of publication in eighteenth century sciences, in contrast to modern science and its emphasis on journal articles. Evidence presented in this chapter suggests that the British were less "professionalized" than the French, so they might have been tied more closely to this older form of publication. Undoubtedly, this was true to some extent, but the evidence shows that the difference was not very great. A reading of Partington (1962) and other bibliographical sources does not indicate that significantly more books were published by British than by French scientists. Further, we know from Chapter 2 (p. 26) and Schofield (1970, p. 191) that some influential texts utilized in Britain were written by French authors. There were a few notable British chemists who did publish books and who are not included in our sample (for example, Lewis, an industrial chemist (Sivin, 1962), and MacBride), but they did not receive many citations (see Chapter 6) and their numbers do not account for the observed great differences. Therefore, it is appropriate to turn to a brief overview of the outlets for communication, the journals.[6]

PERIODICAL LITERATURE

Because the latter third of the eighteenth century was a period of great growth in chemistry, we shall look at the distribution of journals over time to determine how they reflect that growth and to see if there were differences between countries.

Table 3.3 shows that the journals did not exhibit the same magnitude of growth as did the authors and articles. Aside from the period 1760-65, in which there were relatively fewer journals, there was marked growth among journals containing articles only in Britain, from 2 to 6 (with slight growth in Germany —

[5]This explanation could be tested in a crude and preliminary manner by looking at sciences other than chemistry to ensure that the difference was not only in chemistry. From a reading of the historical literature about chemistry and its great attractiveness and usefulness at this time (Gillispie, 1957b; Guerlac, 1959b; Hufbauer, 1971; Langer, 1972; Partington, 1962), one can assume that the situation would be even worse — that is, the differences greater — for other sciences.

[6]Appendix B provides a detailed consideration of the character and contents of the maor journals.

TABLE 3.3
Journals Containing Original Articles[a]

	1760 -65	1766 -71	1772 -77	1778 -80	1781 -84	1785 -86	1787 -88	1789 -90	1791 -95
French	4	7	8	7	8	8	6	7	7
British	2	2	2	2	4	5	5	5	6

[a]For comparison, the equivalent German journals:

	8	11	11	10	11	7	7	9	14.

11 to 14 — only after 1790). The number of French journals fluctuated slightly but exhibited no pattern of growth. I suggest that this was due to the changing character of journals during this period,[7] so that with the establishment of technical (as contrasted with intellectual) journals, either of a general scientific nature such as *Observations sur la Physique,* or specialized nature, such as *Annales de Chimie,* chemical literature moved out of minor and peripheral journals into the major ones. This brought about some consolidation, at least temporarily.[8] In Britain, on the other hand, where few journals of any kind existed in the 1760s and none that specialized until 1797, the growth was largely a reflection of such newly-established scientific organizations as the societies at Manchester and Dublin (cf. McKie, 1948a,b). This is not to say that such societies were not being established in France or Germany — they were — but that the increase from this source in these countries was countered by the phenomenon of consolidation previously noted. In fact, in France some provincial societies which had been established earlier in the century either stopped publishing before or during this period (e.g., Dijon, which stopped in 1786), or did not publish scientific papers at all.

It is worth noting that the difference in the number of French and British journals is less than the difference in the number of articles produced — a 4-to-1 ratio (1772-1777) versus a 6-to-1 — indicating that the reason for the magnitude of the difference previously observed was not merely the absence of places to publish. Moreover, when we see that it was far easier to publish in the major British journal, the *Philosophical Transactions,* than in the *Mémoires* of the Paris Académie, it seems clear that British scientific organization did not promote publication, at least in periodicals. Unlike France (and Germany) there was

---

[7]Compare Hufbauer (1971, Chapter 4) for similar explanations concerning the German case. While the number of German journals is greater than the number of French journals, this is primarily due to the fragmentation of Germany in the eighteenth century. Most of these journals contain only infrequent chemical articles.

[8]The distribution of articles over time in the *Journal de Médecine,* shown in Table 3.4, is a good example of this pattern.

little impetus toward the establishment of journals until very late in the period, and even then publication was infrequent.

Even in France, the vast majority of published articles appeared in only three journals, the Paris *Mémoires,* Rozier's *Observations sur la Physique,* and the *Annales de Chimie* (the last despite the fact that it only began in 1789). While the *Journal de Médecine* was available throughout the period (1754-1794), chemists stopped using it shortly after Rozier's *Observations* took hold; and the *Savans Étrangers,* which provided an opportunity for nonacademicians to publish, also stopped publication as other journals began (cf. Appendix B).

Table 3.4 shows the distribution of articles over time. Since all but three journals had very few articles to distribute over the nine time periods, we can expect minor and insignificant differences in distribution. Beginning with the French journals, we note that *Observations sur la Physique* jumped from a slow start to over 9 articles/year in 1772-77, reflecting its position as the major pub-

TABLE 3.4
Distribution of Articles over Time in Selected Journals,
Cited Sample

	1760 -65	1766 -71	1772 -77	1778 -80	1781 -84	1785 -86	1787 -88	1789 -90	1791 -95
					France				
Obs. Phys.	$-^a$	2	55	24	22	31	14	19	29
N/year	−	2.0	9.2	8.0	5.5	15.5	7.0	9.5	5.8
Méms, Paris	8	16	15	32	41	25	31	7	7
N/year	1.3	2.7	2.5	10.7	10.3	12.5	15.5	3.5	1.4
Ann. Chim	−	−	−	−	−	−	−	37	56
N/year	−	−	−	−	−	−	−	18.5	11.2
Jour. Méd.	12	4	13	1	1	1	0	0	1
N/year	2.0	0.7	2.2	0.3	0.3	0.5	0.0	0.0	0.2
Savans Etr.	7	6	4	10	−	8	−	−	−
N/year	1.2	1.0	0.7	3.3	−	4.0	−	−	−
Méms, Dijon	−	1	1	−	20	6	−	−	−
N/year	−	0.3	0.2	−	5.0	3.0	−	−	−
					Britain				
Obs. Phys.	−	1	4	2	1	5	5	4	0
N/year	−	1.0	0.7	0.7	0.3	2.5	2.5	2.0	0.0
Phil. Trans.	2	9	10	3	13	3	9	9	7
N/year	0.3	1.5	1.7	1.0	3.3	1.5	4.5	4.5	1.4

^{a}− indicates journal not published. The first row for each journal is the total number of articles published over that time period and the second row for each is the mean number of articles per year.

lishing organ of pneumatic chemistry. However, once Lavoisier began seriously to question the phlogiston theory (in *Mémoires,* Paris, for the year 1777, *published* in 1780), the *Mémoires* gained the lead, as Academicians debated chemical issues in its pages, although they also published important articles more quickly in the *Observations* as we have seen in Chapter 2. Once the new paradigm achieved superiority (after 1786), however, the *Observations* lost much of its ground. The establishment of *Annales de Chimie* (1789) appears to have drained the *Mémoires* of much of its talent, but the French Revolution and the slow publication process of the Académie also contributed. Notice that *Journal de Médecine,* which published a substantial number of articles early in the 1760s and 1770s, declined as a chemical journal when competition developed. The publication of *Savans Étrangers* was too sporadic to be conclusive.

As for British articles, we see that *Observations sur le Physique* was a fairly strong competitor of *Philosophical Transactions* in some years, but was not published in as consistently. The articles in the *Philosophical Transactions* exhibit the same pattern as British articles, in general, rising sharply in number in 1766-71 and peaking in 1787-90 at a level of 4.5 articles per year, indicating that the water-nitric acid dispute was more important for British productivity in those years than the inception of the Manchester *Memoirs,* which published a total of only five articles.

Finally, we wish to consider a question raised earlier (p. 17), whether the growth in chemists or in chemical literature was at the expense of other sciences. While we can only infer the growth in chemists, we have evidence for the latter from the three major journals which published over most of the period under consideration: the percentage of pages devoted to chemistry over time. These data can be found in Table 3.5.

The distribution of the percentage that is chemical is similar to the distribution of numbers of authors and articles over time: it shows evidence of growth, albeit with fluctuations, at the expense of at least some other sciences. The journal established last, *Observations sur le Physique,* devoted a larger fraction of its pages to chemistry than the other two, contributing to the sharp rise in chemical production noted in the discussions of Tables 3.1 and 3.2. Furthermore, these are conservative figures, since we have relied on the classifications of the journal editors for the *Mémoirs* and the *Observations:* while all articles they classified as chemistry were included in our sample of articles, so too were some they classified differently, notably some of Lavoisier's and Priestley's articles on "airs."

The results thus far have demonstrated that not only was there a chemical revolution, accompanied by increases in production and productivity as predicted by the theory, but also that there was a gulf between the level of operation of the French chemists and that of the British. Therefore, we must examine more closely the factors which produced this gap. Keeping in mind that social conditions in the eighteenth century have not been completely documented, the primary factors which account for French-British differences are: (1) differences in their scientific organization, (2) differences in their educational structure, (3)

TABLE 3.5
Pages Devoted to Chemistry in the
Three Major Journals

	1760 -65	1766 -71	1772 -77	1778 -80	1781 -84	1785 -86	1787 -88	1789 -90	1791 -95	Total
Mémoires, Paris										
Total	4586	3994	4096	2675	2659	1464	2070	616	788	22938
Chemical	229	223	288	355	467	213	286	48	134	2244
% chem	5.0	5.6	7.0	13.3	17.6	14.5	13.8	7.8	17.2	9.8
Observations sur la Physique										
Total	–	697	5683	3529	3925	1928	1924	1920	480	20086
Chemical	–	71	1019	617	786	487	524	543	118	4165
% chem	–	10.2	17.9	17.5	20.0	25.3	27.2	28.3	24.6	20.7
Philosophical Transactions										
Total	2607	2705	3832	3531	2031	1033	935	968	2170	19812
Chemical	32	222	226	160	326	49	141	118	175	1449
% chem	1.2	8.2	5.9	4.5	16.0	4.7	15.1	12.2	8.1	7.3

NOTE: The first two lines for each journal contain numbers of pages and the third line contains percentages.

differences in the way their governments supported science, and (4) the effects of industrialization. Some consideration of the backgrounds of the chemists of the late eighteenth century will be helpful to us.

BACKGROUNDS AND PRODUCTIVITY

Because backgrounds are not a major part of this investigation, the data are weaker than in other sections and I have relied almost entirely on secondary sources, primarily standard biographical sources (Appendix A: p. 134). Therefore, there is much information available on major chemists and little or none on minor ones.

Although our conclusions are necessarily tenuous, there are still some apparent differences between the two distributions shown in Table 3.6.[9] Of particular note are the large differences between the percentages in "medicine," "pharmacy," and "other science." The 34% difference (20% for France and 54% for Britain) in men connected with medicine may well reflect the greater professionalism and modernity of French chemistry: men who wrote chemical articles in Britain

[9]The percentages are crude and unstable, since more information could increase the total in any given category. However, the results seem to be comparable to the impression one gets in reading histories of the period (Hufbauer, 1971; Partington, 1962; Trengove, 1965).

TABLE 3.6
Percentage of Cited Authors in
Various Fields, by Country[a]

Category	France	Britain
Chemistry	38	42
Medicine	20	54
Pharmacy	29	8
Mineralogy	9	8
Other Science	21	12
Other	23	19
Total number found	66	26
Authors missing	27	1

[a]The table is based only on cited authors. The percentages are based on "Total number found." The categories are not mutually exclusive, so the percentages within each country sum to more than 100.

were overwhelmingly doctors, a profession tied to an earlier age when chemistry was merely the handmaiden of medicine. This accent in British chemistry — which reflects the fact that practically the only training connected with chemistry available in Britain at this time was medical (e.g., at Edinburgh) — is manifested in the very large number of articles written by medical men published in the *Philosophical Transactions* in the years 1740-60, as well as in the period under study. One could argue, however, that a large part of this difference was compensated in the French case by the 21% greater representation of "pharmacy" (29% for France to 8% for Britain).[10] To some extent this is true. However, in France a background in pharmacy seems to have led to a more "scientific" orientation to chemistry than a medical background (although the two were often connected). Some of France's most notable chemists had backgrounds and positions as pharmacists or apothecaries, such as Rouelle and Macquer. Finally, the 9% difference in "other science" (21% for France to 12% for Britain) also reflects a stronger association of chemistry with science in France than in Britain.

One thing not indicated by these data, but clear nonetheless from historical accounts, is the sharp contrast between France and Britain[11] in career opportunities. Despite the fact that there were men in both countries who considered themselves chemists, their careers tended to be different. British scientists were

[10]The strong pharmaceutical connection with chemistry in France and other continental countries, and the absence of this connection in Britain, has been noted also by Guerlac (1961b, p. 74) and Multhauf (1966, p. 272).

[11]And other countries as well: Hufbauer (1971, Chapters 1, 2) shows that Germany also provided few career opportunities in chemistry until very late in the century.

amateurs, both in their approach (Trengove, 1965, p. 184) and in the literal sense: they were not employed as chemists or in related fields (other than medicine), but supported themselves to a large degree, as did Cavendish and Priestley,

Britain also had fairly strong ties to industrial interests, especially in Scotland (Clow & Clow, 1952; Cochrane, 1957; Gibbs, 1952b; Musson & Robinson, 1961; Robinson, 1954; Schofield, 1957b, 1959, 1966). Chemical enterprise in Britain was very closely linked to manufacturing and entreprenuership (esp. Schofield, 1959). Guerlac points out (1959b, p. 96) that Jars, a contemporary of Lavoisier, suggested that, other than trade and manufacturing, the English had few means of gaining wealth. Schofield's extensive studies of the Lunar Society have also made abundantly clear the close association between scientists and industry. The concern with applications and profit perhaps precluded the development of professional norms, values, and career patterns. Industrialization began earlier in Britain than in other countries, and the concomitant rise of the working class may have seriously hindered the development of science even as it heralded the triumph of technology.[12]

French chemists, on the other hand, were often "quasi-professionals" (Hahn, 1971; Multhauf, 1962): they were employed (usually by the government, sometimes by wealthy patrons) as chemists or in related professions such as pharmacy Gillispie, 1957). The active support of the government was in strong contrast to the situation in Britain; this held especially true for the *major* chemists. Of crucial importance in distinguishing the interaction of science and technology in France from that in Britain is the fact that in France distinguished *scientists* were given posts as directors or supervisors of technological or even industrial enterprises, whereas in Britain entrepreneurs and technologists practiced applied science. Gillispie (1957) makes clear that basic science and theoretical ideas were not particularly useful in solving technological problems, but what is important for science is that facilities were provided, rewards were given, and recognition was accorded — all ingredients of successful scientific careers. Furthermore, the Académie Royale des Sciences of Paris, although strictly limited in number, provided stipends and facilities for senior scientists and encouraged a sense of professionalism among its members and aspirants (Hahn, 1971, pp. 86ff).

Among the major differences between the French and British scientific communities was the structure of formal organizations. French science was centered in Paris at the Académie des Sciences: it consisted of a restricted membership (see below) based on scientific performance and provided stipends for senior members. Election was highly competitive and occurred only when a member died. Due to the growing scientific population, the election process had its

---

[12]Compare Gillispie (1957, pp. 402-403) on the different approach and effects of the association of science with technology in France and Britain. He emphasizes the philosophical point of view while we focus on structural aspects.

drawbacks (Hahn, 1971, p. 79), but it generally resulted in a superior level of scientific performance, as evidenced by the productivity of Academicians (shown in Table 3.9) and the high level of recognition accorded them (Chapter 6). This "academy" model was repeated in the provinces, for example, in Dijon, Montpellier, and Rouen.

British science, on the other hand, was neither as centralized nor selective: London's Royal Society, the dominant organization, was relatively noncompetitive, and many amateurs with little scientific competence became members.[13] As a result, the reputation of the Royal Society was at a low ebb (Trengove, 1965; also many contemporary accounts).

Accompanying this organizational difference were two related factors: government provisions of jobs for scientists in France (Gillispie, 1975; Hahn, 1971; Hufbauer, 1971) but not in Britain and a lack of educational facilities for science in Britain (Hans, 1951; Robinson, 1954) but not in France (Taton, 1964). The universities that did exist in Britain were weak in science, with the possible exception of Edinburgh, which came of age only late in the period (Levere, 1970; Morrell, 1971; Robinson, 1954, 1955, 1957, 1958). Furthermore, in Paris there were other facilities, such as the aforementioned Jardin du Roi, where Lavoisier, among others, got a strong chemical background (Guerlac, 1956). Thus there were good opportunities for young men to acquire scientific training in Paris and then to work with established scientists, with Lavoisier at the Arsenal, for example, or even to get government-supported posts themselves. If they were talented, they might get elected to the Académie, become pensioned, and acquire other facilities and sources of income. All this was next to impossible in Britain, and it must have been discouraging to potential chemists (and scientists in general).

As a further test of this theory, let us consider the entry of new authors into the chemical communities in terms of their numbers and productivity. Table 3.7 provides the relevant data. All authors for whom we found evidence of having been published before 1760 in periodical literature were omitted from the table. For the French this involved studying the *Mémoires* of the Paris Académie, *Savans Étrangers,* and the *Journal de Médecine,* and for the British the *Philo-*

---

[13]The relative portions of chemists who were members of these two organizations are shown below. This evidence and reasoning challenge some of the conventional wisdom in the field (sociology of science). Practitioners since Merton (1949) and Parsons (1951) have assumed with no evidence (except Nazi Germany) that science thrives only in a democracy, presumably because it is "democratically" organized — consisting of peer relationships and no appeals to authority. Not only does our general theoretical model, particularly the discussion of factors leading to acceptance or resistance to new paradigms, assume a very different sort of relationship, but the evidence from the organization of French science presented here, as well as the even stronger evidence from the organization of education and science under Napoleon, casts serious doubt on this traditional assumption. Science clearly thrives in a nondemocratic state. It also thrives with strong government support, as we have seen again in recent times.

TABLE 3.7
New Authors and Their Articles

	1760 -65	1766 -71	1772 -77	1778 -80	1781 -84	1785 -86	1787 -88	1789 -90	1791 -95
France									
New authors	11	7	24	13	11	6	4	5	5
Authors added per year	1.8	1.2	4.0	4.3	2.8	3.0	2.0	2.5	1.0
Articles[a]	18.0	8.0	30.3	16.5	21.0	8.5	5.3	6.3	11.0
% of total FR[b]	67	28	33	23	23	11	10	10	10
Britain									
New authors	1	5	5	1	3	0	2	1	5
Authors added per year	0.2	0.8	0.8	0.3	0.8	0.0	1.0	0.5	1.0
Articles[a]	1	8	8	1	5	0	4	1	7
% of total BR	50	80	53	20	36	0	29	6	54

[a] Articles by men who have been publishing for no more than two years.

[b] FR = French, BR = British. These figures were obtained by eliminating all authors for whom evidence could be found of having been published in *periodical* literature before 1760.

sophical *Transactions* of the Royal Society, London, and the *Medical Commentaries*. . . at Edinburgh.

Table 3.7 presents an interesting and mixed picture. The number of new authors entering the field of chemistry in each period makes it abundantly clear that there was relatively little recruiting of chemists in Britain until late in the century (1791-95). Both groups showed a high rate of growth (relative, that is, to their respective rates at other times) at approximately the time that pneumatic chemistry was being established: 1766-71 in Britain (0.8 authors/year) and 1772-77 in France (4.0 authors/year). However, since the number of British authors was much smaller than the number of French authors, the fraction of articles contributed by this new group was higher for Britain than for France. For the number of new authors added per year, there was a peak in France in the 1770s, probably reflecting the sudden interest in pneumatic chemistry and the opportunity for publication offered by the *Observations sur la Physique*. . . .

Looking at the proportion of articles contributed by new authors (authors who had been publishing for no more than two years), we find that for the first 18 years (1760-77), British newcomers wrote more than half of all British articles. Although this was followed by periods in which the proportion of articles written by new members of the community was lower, we find that in 1778-80 the one new author wrote 20% (or one) of the articles and that this was followed by a period in which new authors wrote over a third of all British articles. Aside from the 67% in 1760-65, the proportion of French articles contributed by new authors was generally lower. As the French community grew throughout the latter part of the century, the proportion of articles by new authors declined to a stable rate of 10% from 1787 on. The difference between the two countries again indicates the greater professionalism of the French chemists: they were an established group of authors whose new members could only gradually achieve full membership, in contrast to Britain, where new authors practically dominated, especially in the 1760s and 70s.

While the pattern for Britain would seem to be disastrous for a period of normal science — when those least experienced in using the paradigm produced most of the literature — perhaps it is helpful in the case of a revolution. In fact, in line with the reasoning developed in Chapter 1 (pp. 18-19), we should expect that this condition would lead to a greater receptivity of a new paradigm. However, we know that British chemists were more resistant to change than French chemists. An obvious and plausible explanation, common to historical literature, is that this was an historical accident: the new paradigm happened to have been invented by an unusual chemist who happened to be French, and nationalism (or other idiosyncratic factors) did the rest. We would prefer, however, explanations which are not ad hoc. Since we have seen that productivity in Britain was very low, and that British chemistry (and science in general) was at a low ebb, providing little incentive to new members and attracting relatively few, we might wonder about the nature of the new members. The attribute that relates most pertinently to our theory of which scientists are most resistant to change is age. Table 3.8 gives the average age of authors at the time of their first publication.[14]

As we might expect, new British authors were older than new French authors in all but two time periods (in 1766-71, when the British took the lead in pneumatic chemistry, and in 1791-95, when France was in political upheaval). However, average ages do not tell the whole story. First of all, both sets of figures seem rather high except for the French from 1781 to 1786, the peak period of debate, and from 1789 to 1790. Looking at the overall distribution more closely, we find that almost a third (20 or 31%) of French authors were 30 or younger when they published their first chemical article, and over half (34 or 52%) were 35 or younger. Of British authors, on the other hand, only 13% (3) were 30 or younger and less than a third (7 or 30%) were 35 or younger. At the extremes, nine French (14%) and only one British author were 25 or younger, while ten French (15%) and six British (26%) were 50 or older when they published their first chemical article. Quite clearly, then, a larger fraction of French than of British chemists began their careers at a young age. The reason that the average

---

[14]This is the mean age of *authors* and should not be compared with the ages reported in Chapters 4 and 5 which are ages of *article-writers:* that is, here the unit of analysis is an author, and the age of an author is included only once, whereas in later chapters the unit of analysis is an article, and the age of an author is included for every article he wrote.

TABLE 3.8
Average Age of Authors at Publication of First Article

	1760 -65	1766 -71	1772 -77	1778 -80	1781 -84	1785 -86	1787 -88	1789 -90	1791 -95	Total
French	36	41	38	38	32	32	43	30	37	36
N	8	7	17	7	10	5	4	4	3	65
British	55	38	42	41	50	–	46	39	34	41
N	1	5	5	1	3	0	2	1	5	23

Note. The table is restricted to authors who were cited. N = the number of men for whom ages were calculated. See the note to the previous table for the selection procedure for "new" authors.

for the French authors is not much lower than that of the British is that in France the Revolution or the recruitment process attracted older men *from other fields* as well as new scientists, a phenomenon we predicted when we derived our productivity hypotheses (see pp. 17-19). The British authors do not appear to be established scientists in other fields who were being drawn to chemistry late in life,[15] although many of them were medical doctors. Even the most notable British chemists, Priestley (39), Kirwan (48), and Cavendish (35), got their start later than their French counterparts, Lavoisier (25), Guyton (32), Berthollet (28), and Fourcroy (30).

This result appears to shed light on two issues. First, it confirms the discussion of the quality and availability of training and career opportunities for French scientists, at least in chemistry: since training was more readily available, men could start their careers earlier. Second, and more important, it helps explain the greater reluctance of British chemists to accept the new theory; in spite of the fact that they were newer to the field, on average, than French chemists, they were also older and therefore less receptive to the new ideas, as will be shown in Chapters 4 and 5. Although they may not have published very widely (at least in journals – a point to be discussed below), they had lived a longer time under the influence of the phlogiston paradigm and had spent their youth under its dominion.[16]

[15] It would be nice to be able to separate out the two groups, but due to lack of information in many cases, this is difficult and risky. It is clear only for Academicians, such as Tillet (from botany at 52) or Haüy (a crystallographer, from botany at 50). Even Lavoisier's close associate, Laplace, who was elected to the Académie in mechanics at age 24 (1773) did not contribute to chemical literature until he was 35.

[16] As an example, consider the average "new" chemist in Britain in 1781-84. He was 20 in 1751-54 and 30 in 1761-64, the period during which the phlogiston paradigm became dominant in Britain. On the other hand, the average "new" French chemist in 1781-84 was only 35. Thus, he was 20 in 1766-69 and 30 in 1776-79, the period in which phlogiston was coming under attack. Furthermore, since the new French chemists were even younger after this time, on average, there would be almost no phlogistic influence on them.

Unfortunately, we cannot rely very heavily on these data since they are incomplete for the French. It is conceivable that the 30% of French authors whose biographies could not be found were significantly older than those whose could. The argument that those who got started early would be more likely to establish themselves and thus become more noticeable to biographers seems reasonable. On the other hand, we know that the average age for new French chemists was inflated in some periods by older men from other fields trying their hand at a new game (Brisson was an example). Therefore, although the data are suggestive, and are supported by the results of Chapters 4 and 5, they are not conclusive.

We have already mentioned the differences in opportunity between French and British chemists and the importance of the Académie des Sciences in Paris. We now turn to a further consideration of the relative importance of this organization and its closest British counterpart, the Royal Society of London. We have already found that membership in the Académie was restricted while that of the Royal Society was not, and that the Académie provided support for its senior members while the Royal Society did not. Table 3.9 shows their productivity in relation to that of their colleagues. We are following the example of Hahn (1971, p. 97) in the table, restricting it to "working" Academicians: those who were regular members of one of the sections and neither "honoraire" nor "libré" (Hahn, 1971, pp. 77-78).

Of the 102 French authors in our sample, 25% (26) were regular Academicians, 15% (16) in chemistry, and the other 10% (10) in other sections.[17] These men produced 46% (283) of the articles in the cited sample and, even more strikingly, the 16 chemical members produced 40% (244).

Looking at the distributions over time, we see that the Academicians were overrepresented in number (relative to their overall proportion of 25%) in every period except 1760-65, indicating that they published more regularly than their compatriots. They also contributed a disproportionate share of the articles in every period except 1760-65.[18] They were particularly dominant in terms of numbers just before the appearance of the *Observations sur la Physique* (43% in

[17]If we consider the 93 cited authors as our base, then 17% were members in chemistry and 28% were regular members. Another five men were "libré" or "honoraire," and an additional six (three in chemistry) were elected regular members in 1795, so that a total of 37 of the 93 cited authors (40%) were connected with the Académie at some point, a remarkably high percentage given Hahn's remarks (1971, pp. 97-98) about the difficulties of being elected.

The sixteen chemical members were Baron, Baumé, Berthollet, Bourdelin, Bucquet, Cadet, Cornette, Darcet, Disjonval, Duhamel du Monceau, Fourcroy, Hellot, Lavoisier, Macquer, Pelletier, and Sage. The three chemists elected in 1795 were Bayen, Guyton de Morveau, and Vauquelin; the ten regular, nonchemistry section members were Brisson, J. P. Duhamel, Haüy, Herissant, Laplace, Lassone, Meusnier, Monge, Tillet, and Vandermonde.

[18]That is, the proportion of articles contributed was greater than their proportion of authors publishing in every period (except 1760-65).

TABLE 3.9
Production of Members of the Paris Académie des Sciences
and the Royal Society of London

	1760 -65	1766 -71	1772 -77	1778 -80	1781 -84	1785 -86	1787 -88	1789 -90	1791 -95
France (Académie des Sciences)									
Authors/year	0.7	1.8	2.8	5.3	4.5	5.0	7.5	4.5	3.0
% of total FR[a]	18	43	26	53	38	39	56	30	28
Articles	4	14.5	25.5	39	43	37	35.3	38.3	46.8
Articles/year	0.7	2.4	4.3	13.0	10.8	18.5	17.7	19.2	9.4
% of total FR	15	50	28	54	47	49	69	59	41
Art/auth/year	1.0	1.3	1.5	2.4	2.4	3.7	2.4	4.3	3.1
% of total FR	83	120	107	102	126	128	124	194	149
Britian (Royal Society)									
Authors/year	0.2	1.0	1.3	1.3	1.8	2.0	4.0	5.5	1.2
% of total BR	50	60	67	80	64	100	89	92	55
Articles	1	8	10	4	9	8	14	18	9
Articles/year	0.2	1.3	1.7	1.3	2.3	4.0	7.0	9.0	1.8
% of total BR	50	80	67	80	64	100	93	95	69
Art/auth/year	1.0	1.3	1.3	1.0	1.3	2.0	1.8	1.6	1.5
% of total BR	100	133	100	100	102	100	105	104	127

[a]FR = French, BR = British. Authors/year are the average number of authors publishing per year; articles/year are the average number of articles published per year; and art/auth/year are the average number of articles published per year per author publishing (authors/year divided into articles/year). The percentages are of the appropriate rate of something per year for the entire "cited" sample (from Table 3.2). The data for the French are restricted to "working" members, that is, honorary and "libré" members were excluded.

1766-71), when Lavoisier began to seriously question phlogiston theory (53% in 1778-80) and as conversions to the new theory spread rapidly (56% in 1787-88).[19] They were particularly dominant in articles published throughout the period of debate (1778-90), peaking again in 1787-88 at 69%. Furthermore, although the proportion of authors who were Academicians declined after 1788, and the proportion of articles contributed by them declined after 1790, this was the period of their greatest relative productivity: in terms of articles per author per year they produced at a rate 94 and 49% greater than the average French chemist (including Academicians) in 1789-90 and 1791-95, respectively. Thus, during the period when the French Revolution was constricting science and the Académie

[19]See Chapter 4. Recall that the *Mémoires* were printed two or three years in arrears, so that the timing of articles published in it was delayed.

itself was suppressed, Academicians more than held their own (compare Crosland, 1967, pp. 174-175).

Since the early converts to the new theory were Academicians (Lavoisier, Berthollet, Fourcroy, Laplace), we see that the Académie was instrumental in establishing the new paradigm. Given its prestige, this was undoubtedly an important factor in the swift acceptance of the new theory in France. Aspiring young French chemists had little choice but to accept the new theory.

In Britain the situation was quite different. Since election to the Royal Society was so easy, essentially a matter of desire, fully 78% (21) of British authors were Fellows. However, they only published a slightly greater proportion of articles, 80% (81), than their representation would warrant. Looking at the data over time, we see that they were overrepresented in number in only three time periods, 100, 89, and 92% in 1785-86, 1787-88, and 1789-90, respectively. In the proportion of articles which they contributed, they were overrepresented (relative to their proportion of authors) in four periods, 1766-71 (80 to 60%), 1787-88 (93 to 89%), 1789-90 (95 to 92%), and 1791-95 (69 to 55%). However, in only three periods, 1785-86, 1787-88, and 1789-90, were both their relative number and relative number of articles greater than their overall representation among British chemical authors, and in no case was the difference statistically significant. The representation of Fellows, then, was at its peak around the time of the greatest conflict over oxygen theory, in 1785-86, when the French began to convert, and in 1787-90, when the debate over water and nitric acid erupted.

We shall return to further consideration of the influence of these organizations in the discussion of citation data in Chapter 6.

SUMMARY

This chapter begins to show some of the many differences between the chemical communities of France and Great Britain and the relationship of these differences to the Chemical Revolution. The first important finding was that the French chemists were both more numerous and more productive than the British chemists during this time period (1760-95). The change in their number and productivity over time followed closely our predictions, (H8) and (H9), that there would be rapid growth of chemists and articles after 1777 (or possibly 1772, because of the widespread enthusiasm for the new chemistry of gases).

There were some differences in background, but of far greater importance was the finding that new recruits to the field were fewer and older in Britain; we attributed this to the lack of good training grounds and/or stable career opportunities. We found that the Paris Académie des Sciences dominated French chemistry far more than its restrictive requirements and membership numbers would indicate, both in articles produced and articles per author.

The dominance of French chemists was attributed to France's superior scientific organization and two characteristics were particularly important: (1) there were more opportunities for both training and professional careers in France; and (2) closely related to this, there was a strong, centralized organization of scientists, at the head of which was the Académie des Sciences of Paris, supported by the government. These factors combined to produce a better trained and more professional group of chemists in France than in Britain.

These results are in contrast to the assertions of Hahn (1971, p. 137) regarding the relative merits of French and British scientific organization. Hahn implies that British science was flourishing, except for the admittedly moribund Oxford and Cambridge, and that the Royal Society and other British organizations were viewed with admiration by the French. Not only is his view contradicted by the data here, but it is also questionable in view of contemporary — though perhaps not modern — accounts.[20] In his eagerness to explain that attacks on the Académie during the French Revolution were due to the rigidity of that organization, he appears to have engaged in overkill, particularly in view of the fact that even revolutionaries came under the guillotine during that period.

Hahn uses the observation that the mean age at election increased monotonically over the century from the 1730s through the 1780s (1971, p. 97) as part of his evidence to explain the pernicious effects of the regulations of the Académie regarding its fixed size. However, in the footnote in which he gives the values he calculated, there are only five numbers for six decades. Furthermore, most unfortunately, he has picked a starting date especially suited to his point: the 1730s were characterized by the lowest mean age of election of any decade in the century, due in part to the election, by *special* dispensation, of Clairaut at the age of 18, which Hahn mentions, but implies that it was not atypical. If we calculate the mean age at election of "working" members of the Académie, (for the 140 of 156 members elected in the eighteenth century for which birthdates are given in the *Index Biographique*) we get the following results: 36.5 (including an *élève* of 15), 34.9, 32.4, 27.9, 31.4, 31.4, 31.7, 37.1, 37.1 for the decades beginning 1700, 1710, . . ., 1780, respectively. Thus it was not until the 1770s that the mean age exceeded the level it stood at in the decade of 1700-10, and only then because of a geologist elected (to the only position in that field) in 1773 at age 75! If we exclude him, the average for 1770-79 is only 35.0. An F-test shows no significant differences among decades.

Thus we get a picture in France of a professional (or quasi-professional) scientific community: there were centers for training, career possibilities, outlets for publication, and rewards for good performances, including election to the

[20]Compare the discussion on pp. 38-39 above. See also Chaldecott (1968) and Crosland (1969).

prestigious Académie. Chemists and other scientists were encouraged to publish in journals — that is, to contribute information to the community, relatively rapidly, and they were recognized for their contributions (see Chapter 6) and supported for their work.

This is in sharp contrast to Britain, where scientists were still amateurs. Britain had no strong organization, no state support of research, and very little support of any other kind. The evidence presented shows that there was little recruitment of new members for the chemical community and little opportunity for them to make contributions.

4

Conversion and Resistance

Now that we have sketched a broad picture of the structure of the chemical communities of France and Great Britain which produced the Chemical Revolution, we are ready to look at the development of that revolution in detail. We shall specifically examine the characteristics of the chemists and the literature that accompany such radical change.

BASIC DISTRIBUTIONS[1]

We begin by looking at the change in "attitudes" toward the two major paradigms over time.

Table 4.1 shows that four of the hypotheses are supported by this set of data. First (H1): the proportion of oxygen articles will increase and the proportion of phlogiston articles decrease over time. For French articles the proportion oxygen jumps from 0 to 66% (33 "oxygen" and 33 "new nomenclature") while the proportion phlogiston fluctuates from 26 to 52 to 7% (4 "phlogiston" and 3 "old nomenclature"). For British articles, the proportion oxygen rises from 0 to 16%, while the proportion phlogiston varies from 10 to 88 to 15%. Obviously, the proportion phlogiston among British articles did not decrease smoothly over the whole period. In fact, it increased for the first 25 years. This does not indicate that the hypothesis is wrong, however, merely that it must be restricted to the proper time period: once the new theory becomes a serious challenge (1785-86), the proportion phlogiston does decline, but not as much as among French articles. The initial rise in the proportion phlogiston may be considered a reaction by the supporters of the old paradigm to the attack by those of the new: they

[1]The distributions in this section are based on the sample of *cited* authors (see Appendix A for descriptions of samples) consisting of 616 French and 101 British articles. The equivalent distributions for the "basic" sample are given in Appendix C.

TABLE 4.1
Percentage Distribution of Articles over Time by
Type of Theory, Controlling for Country (Cited Sample)

	1760 -71	1772 -77	1778 -80	1781 -84	1785 -86	1787 -88	1789 -90	1791 -95	Total
					France				
Phlogiston	39	49	38	42	29	14	9	4	28
Old nomenclature after 1787	_[a]	–	–	–	–	–	9	3	1.5
Neutral	2	–	–	–	3	–	–	–	0.5
Antiphlogiston	–	3	11	2	8	4	5	–	4
New nomenclature	–	–	–	–	–	8	34	33	10
Oxygen	–	–	–	8	19	33	29	33	15
No information	59	48	51	48	41	41	14	27	41
Total	100	100	100	100	100	100	100	100	100
Number of cases	(56)	(92)	(72)	(91)	(75)	(51)	(65)	(114)	(616)
					Britain				
Phlogiston	9	40	20	71	88	40	58	15	44
Old nomenclature after 1787	–	–	–	–	–	27	5	31	9
Neutral	–	–	–	–	–	–	5	–	1
Antiphlogiston	–	–	–	–	–	–	–	–	–
New nomenclature	–	–	–	–	–	7	16	8	5
Oxygen	–	–	–	–	–	–	–	8	1
No information	91	60	80	29	12	27	16	39	40
Total	100	100	100	100	100	101[b]	100	101[b]	100
Number of cases	(12)	(15)	(5)	(14)	(8)	(15)	(19)	(13)	(101)

[a] – indicates no articles.

[b] Rounding error.

are forced to defend themselves, thus stressing their basic theoretical concept in a larger proportion of their papers.[2]

The increase in the oxygen categories, however, began earlier and was markedly greater for French articles, supporting (H5) and (H11). The fact that there are no British articles classified "antiphlogiston" is probably due to the fact that by the time any British authors seriously questioned the old paradigm both the new nomenclature and the term (and concept of) "oxygen" were available.[3]

[2] See Chapter 1, p. 16 and compare Kuhn, (1952, p. 14; n 13). The effect was mitigated among French articles due to the publications of Lavoisier and the relatively high proportion of "No information" from 1781 to 1788.

[3] Although this does not completely rule out "antiphlogiston," as can be seen from the existence of such articles in France after 1787, it does make it unlikely.

Third, the table shows that (H10) is correct: there were very few "neutral" articles; this fits Kuhn's "gestalt switch" analogy very well and casts doubt upon "incremental" or "conjecture-refutation" views of scientific revolutions (see Chapter 1, pp. 19-20).

At first glance, the table seems to show that phlogiston, though a paradigm, was not a very widespread one, for the proportion of articles following it exceeded 50% in only one period among French articles and three periods among British, and in these latter periods only after the initial challenge by antiphlogiston forces, 1772-1780. When we realize, however, that phlogiston, despite its many properties,[4] was not relevant for many reactions, this result is not surprising.[5]

The final noteworthy feature of the data shown in the table is the distribution of "No information" articles, which constitute 40% of all articles published. There were relatively more of them during the early part of the period, and the proportion of articles so-classified declined with the coming of the controversy (especially in Britain) and increased slightly with the triumph of the new paradigm.[6] It appears that the revolution drew out opinions and/or focused research on relevant issues. While this is certainly plausible, the evidence is inconclusive – the extreme drop after 1788 also reflects the fact that the use of a new nomenclature after it became available (1787) allowed an article to be classified without explicit mention of oxygen or phlogiston, whereas earlier there was no such option.[7]

Given the paradigm switch, we have predicted accompanying changes in the levels of theoretical and quantitative research. Table 4.2 shows that the relation-

---

[4] For example, it not only explained calcination and combustion, but also accounted for colors, odors, the metallic properties of metals, and the similarities in the reactions of the known acids. See the discussion of Rouelle's theory in Chapter 2, p. 25).

[5] Another gauge of the extent to which phlogiston dominated during the years 1760-1785 is the proportion of authors who used it. Of all authors who published before 1785, 65% (48 of 74) of the French and 50% (11 of 22) of the British used phlogiston in at least one article. The other authors published only "no information" articles. With rare exceptions, these latter authors published only single articles, so it is difficult to say whether or not they supported phlogiston from the data. In some cases we know from other sources that they did. The question of the paradigmatic nature of phlogiston is taken up again on pp. 120-122.

[6] The rise in France may be due partly to "extraneous" factors, such as the state of publication during the French Revolution when many major journals were suspended.

[7] Although it is highly unlikely that an author would use the new nomenclature if he did not support the new paradigm (and we have assumed this in all cases), he might very well have used it in a case where neither phlogiston nor oxygen was relevant – before the new nomenclature, this would have resulted in a "no information" classification.

Table 4.2
Percentage Distribution of Articles over Time by
Level of Theory, Controlling for Country

	1760 -71	1772 -77	1778 -80	1781 -84	1785 -86	1787 -88	1789 -90	1791 -95	Total
				France					
Descriptive	61	56	47	39	32	43	55	57	49
Mixed	38	37	41	45	49	41	34	34	40
Theoretical	2	7	12	16	19	16	11	9	11
Total	101[a]	100	100	100	100	100	100	100	100
Number of articles	(56)	(92)	(73)	(91)	(75)	(51)	(65)	(114)	(616)
				Britain					
Descriptive	33	47	60	21	12	47	32	61	39
Mixed	59	33	40	43	25	40	53	31	41
Theoretical	8	20	0	36	63	13	16	8	19
Total	100	100	100	100	100	100	101[a]	100	99[a]
Number of articles	(12)	(15)	(5)	(14)	(8)	(15)	(19)	(13)	(101)

[a]Differs from 100 because of rounding.

ship between the chronology and proportion of articles which were theoretical is as predicted by (H2): the proportion of theoretical articles did increase to a peak in 1785-86 and then declined — precipitously in Britain.[8]

Contrary to our Hypothesis (H19) that French chemists would exhibit a higher level of theoretical work than British, the proportion of French articles which was theoretical is clearly lower than the proportion of British articles. The most likely reason for this is that the French were more prone to publish in journals, and, more specifically, that there were many more outlets for applied chemists in France than in Britain. While British applied chemists did have organizations available, such as the Lunar Society, these organizations did not usually publish journals. The French, on the other hand, with their vast, centralized organization in Paris, allowed and even fostered the publication of chemical articles by applied or other nontheoretical chemists. Thus, paradoxically, the professionalism of the French scientific community led to a lower proportion of theoretical articles. This result, however, challenges only the assumptions about

[8]Part of this decline may be due to the effects of the French Revolution. We have seen in Chapter 3 that many journals were suppressed in France and that productivity declined in both France and Britain. However, to attribute this decline (and a decline in many of the distributions that follow) to the French Revolution, one would have to assume that the suppression affected theoretical articles (or, for the distributions that follow, the appropriate category) *disproportionately*. We have no evidence that this was, in fact the case.

TABLE 4.3
Percentage Distribution of Articles over Time by
Level of Quantification, Controlling for Country

	1760 -71	1772 -77	1778 -80	1781 -84	1785 -86	1787 -88	1789 -90	1791 -95	Total
				France					
% Quantitative	19	28	47	27	33	35	31	39	33
Number of articles	(56)	(92)	(73)	(91)	(75)	(51)	(65)	(114)	(616)
				Britain					
% Quantitative	25	0	40	50	63	20	32	46	32
Number of articles	(12)	(15)	(5)	(14)	(8)	(15)	(19)	(13)	(101)

France and Great Britain and the effect of "professionalism," not the theory of revolutions presented.

In Table 4.3, if we compare the proportion quantitative in each period with the average (33% for France and 32% for Britain), the later periods in both countries appear somewhat more quantitative in nature than the earlier periods, as hypothesized (H3), but the differences are not very great. Furthermore, the overall difference between France and Britain is negligible, contrary to Hypothesis (H20). However, the patterns in the two countries are quite different. The French articles showed slow, steady growth in the proportion quantitative, except for the peak 1778-80,[9] whereas the British articles exhibited very low proportions quantitative through the late 1770s, after which a sharp increase occurred, ending in the late 1780s. This would appear to support our position that there was a continental tradition of quantification (see Chapter 2, p. 29).

FURTHER ANALYSIS

The results of the previous analysis are somewhat veiled, for they are based on all articles written by cited authors. For analytical purposes, it seems reasonable to eliminate articles which did not take a position as to which paradigm they support (articles classified as "no information") and that has been done in the following analyses. There are two reasons for this: (1) many of these articles were concerned with topics for which one or another of the paradigms was

[9]This "peak" was caused by a sharp increase in the proportion of oxygen papers, papers which were highly quantitative (expected by (H3)), as can be seen from inspection of Tables 4.6, 4.9, and 4.10 below.

TABLE 4.4
Percentage Distribution of Articles over Time by
Level of Theory, Controlling for Country: Analytic Sample

	1760 -71	1772 -77	1778 -80	1781 -84	1785 -86	1787 -88	1789 -90	1791 -95	Total
				France					
Descriptive	48	26	32	21	14	23	47	46	33
Mixed	48	61	52	49	56	50	40	41	48
Theoretical	5	13	16	30	30	27	13	13	18
Total	101[a]	100	100	100	100	100	100	100	99[a]
Number of articles	(21)	(46)	(31)	(47)	(43)	(30)	(55)	(78)	(351)
				Britain					
Descriptive	0	33	100	10	14	40	20	50	28
Mixed	100	17	0	40	14	40	60	38	40
Theoretical	0	50	0	50	71	20	20	12	33
Total	100	100	100	100	99[a]	100	100	100	101[a]
Number of articles	(1)	(6)	(1)	(10)	(7)	(10)	(15)	(8)	(58)

[a]Differs from 100 because of rounding.

not relevant, and (2) this category does not fit into the ordinal scale used to measure opinion on the competing paradigms.[10]

Table 4.4 shows strong support for Hypothesis (H2). There is a clear rise in the proportion of articles which were theoretical in both countries during the years of crisis and debate, although the small numbers for Britain can only be suggestive.

Elimination of the "no information" articles does result in stronger patterns: in relation to Table 4.2, Table 4.4 shows an increased proportion of theoretical articles for both Britain and France, as well as an increase in the proportion "mixed" for France — a result of the strong association between the "no information and "descriptive" categories (see Appendix A). The distributions are similar, indicating that the reduced sample does not distort the data, but there is even stronger support for the hypothesized curvilinear relationship between time and level of theory: again the proportion theoretical peaked in 1785-86 and then declined sharply. We can explain this decline by noting that at that time there was a new period of normal science with a new characteristic level of

[10]For reasons of consistency and comparability with the results for the shift hypotheses in the next section and those of the analyses of Chapter 5, we have also eliminated all articles by authors for whom we could not find a birthdate. We are, therefore, using the "analytic" sample throughout this section. A fuller discussion of the various samples and the implications of omitting certain authors or articles appears in Appendix A.

TABLE 4.5
Distribution of Quantitative Articles over Time by
Country: Analytic Sample

	1760 -17	1772 -77	1778 -80	1781 -84	1785 -86	1787 -88	1789 -90	1791 -95	Total
					France				
% Quantitative	14	39	45	32	47	53	29	46	39
Number of articles	(21)	(46)	(31)	(47)	(43)	(30)	(55)	(78)	(351)
					Britain				
% Quantitative	100	0	0	50	71	20	33	25	34
Number of articles	(1)	(6)	(1)	(10)	(10)	(10)	(15)	(8)	(58)

theory. This is reasonable for France, but not for Britain, where the Revolution was not yet successful. The small numbers make explanations tenuous, and perhaps unnecessary, but the decline may well be due to the retirement from the debate of unsuccessful phlogiston chemists.[11]

Again, relative to the data shown earlier (Table 4.3), the results are improved. Table 4.5 shows a nonmonotonic increase in the proportion of French articles that were quantitative (from 14% in 1760-61 to 53% in 1787-88), as predicted; but the increase is not as great as we expected. However, after a high point (71%) in Britain in 1785-86 there was a sharp decline, contradicting Hypothesis (H3).

Both of these anomalies may be due to several factors: (1) the presence of new problems for the new paradigm that were not readily amenable to quantification (of the sort we are measuring); (2) the avoidance of quantitative work by phlogiston chemists, which would affect the British results very strongly; and (3) the possibility that quantification was not institutionalized by the Revolution, contradicting Hypothesis (H3) and most secondary literature.[12] Although further detailed study of the articles is necessary to demonstrate this conclusively, we will assume that a combination of the first two factors was the primary cause, consistent with the theory. Factor (2) will be considered further during discussion of the shift hypotheses.

[11]This will be reconsidered below after we have looked at the switch hypotheses and controlled these relationships for paradigm.

[12]There is also the problem of measurement error: the recipes and "pseudoquantification" discussed in Appendix A. For this to affect these results, one must assume that there was more pseudoquantification in earlier periods than in later, which is not unlikely.

SHIFT HYPOTHESES

There are four hypotheses about shifting allegiance: (1) French articles will shift to the oxygen paradigm earlier than British ones; (2) theoretical papers will shift earlier than nontheoretical ones; (3) quantitative papers will shift earlier than qualitative ones; and (4) papers by younger authors will shift earlier than papers by older ones. In this section we will examine these hypotheses and related aspects of the data using the analytic sample.

The first step is to look at the raw distribution of phlogiston and oxygen articles over time, which we present in Table 4.6 (this table is comparable to Table 4.1, but with the categories collapsed, and the "no information" articles and articles with no author age omitted).

Table 4.6 supports Hypothesis (H11), as did Table 4.1: we find in the analytic sample, as in the cited sample, that there was an increase in the proportion of

TABLE 4.6
Percentage Distribution of Articles over Time
by Paradigm, Controlling for Country[a]

	1760 -71	1772 -77	1778 -80	1781 -84	1785 -86	1787 -88	1789 -90	1791 -95	Total
				France					
Phlogiston	95	94	74	81	49	23	22	9	49
Number of articles	(20)	(43)	(23)	(38)	(21)	(7)	(12)	(7)	(171)
No./year	1.7	7.2	7.7	9.5	10.5	3.5	6.0	1.4	4.8
Oxygen	0	7	26	19	47	77	78	91	50
Number of articles	(0)	(3)	(8)	(9)	(20)	(23)	(43)	(71)	(177)
No./year	0	0.5	2.7	2.3	10.0	11.5	21.5	14.2	4.9
				Britain					
Phlogiston	100	100	100	100	100	100	80	38	86
Number of articles	(1)	(6)	(1)	(10)	(7)	(10)	(12)	(3)	(50)
No./year	0.1	1.0	0.3	2.5	3.5	5.0	6.0	0.6	1.3
Oxygen	0	0	0	0	0	0	13	38	9
Number of articles	(0)	(0)	(0)	(0)	(0)	(0)	(2)	(3)	(5)
No./year	0	0	0	0	0	0	1.0	0.6	0.1

[a]Percentages sum down columns within country. The number of neutral articles was so small that they were omitted from the table. Thus, some of the percentages do not add up to 100: among French articles there was one neutral in 1766-71 and two in 1785-86 and among British articles one in 1789-90 and two in 1791-95. Column totals are omitted.

TABLE 4.7
Percentage Distribution of Oxygen Articles over
Time By Level of Theory, Controlling for Country
(Percentage Oxygen or AntiPhlogiston)

	1760 -71	1772 -77	1778 -80	1781 -84	1785 -86	1787 -88	1789 -90	1791 -95
				France				
Nontheoretical	0	0	6	4	44	75	80	89
Number of articles	(16)	(31)	(18)	(23)	(18)[a]	(16)	(44)	(53)
Theoretical	0	20	54	33	48	79	73	96
Number of articles	(5)[a]	(15)	(13)	(24)	(25)[a]	(14)	(11)	(25)
				Britain				
Nontheoretical	0	0	0	0	0	0	25	0
Number of articles	(1)	(2)	(1)	(1)	(1)	(8)	(8)[a]	(4)[b]
Theoretical	0	0	0	0	0	0	0	75
Number of articles	(0)	(4)	(0)	(9)	(6)	(2)	(7)	(4)

[a]Includes one neutral article.   [b]Includes two neutral articles.

oxygen articles among the French beginning in 1772-77 and among the British beginning in 1789-90. There was also a sharp jump in this proportion for French articles in 1785-86, corresponding to the years in which major conversions began, after which the oxygen articles dominated. The number of British phlogiston articles, in contrast, continued to rise until 1791-95, an indication of strong British resistance to the new paradigm.

In a sense, the French phlogiston articles after 1786 were progressively more "anomalous." It would therefore be helpful to know something about the men responsible for them. Rather than repeating their names in several places, we will put off discussion until after we have looked at the remaining shift hypotheses. These hypotheses involve controlling the relationship between time and paradigm choice for other variables: theoretical level, level of quantification, and age. We will then have a better idea of the nature of the defenders of the old paradigm and converts to the new.

We continue the examination of the shift hypotheses by controlling first for theoretical level, which has been dichotomized in Table 4.7.

Table 4.7 shows that among French articles, the hypothesis that the more theoretical would shift first has been supported[13]: the percentage oxygen among

[13]It is possible that these results are due almost entirely to the work of Lavoisier (and being due to one man, not typical of a revolution). However, when all of Lavoisier's articles are subtracted from the table, the conclusions do not change. (See Appendix C, Table C.6.)

theoretical articles was 20, 54, and 33 for the periods 1772-77, 1778-80, and 1781-84, respectively, while for nontheoretical it was only 0, 6, and 4%. Further, the proportion of papers which used oxygen was higher among theoretical than nontheoretical articles in every period except 1789-90. The decline in percentage oxygen among theoretical articles in 1781-84, and the small difference in this proportion for theoretical and nontheoretical articles in the subsequent periods was a result of the theoretical reaction of the phlogiston chemists to the attack of antiphlogistonists, as we predicted.

If we look at the proportion of phlogiston articles which were theoretical, this becomes quite apparent, for the proportion rose strongly in 1781-84 and again in 1785-86, as predicted, after which the *number* of phlogiston articles dropped off rapidly:

TABLE 4.8
Percentage of French Articles which Were Theoretical by Paradigm

	1760 -71	1772 -77	1778 -80	1781 -84	1785 -86	1787 -88	1789 -90	1791 -95	Total
Phlogiston	20	28	26	42	57	43	25	14	33
Number of articles	(20)	(43)	(23)	(38)	(21)	(7)	(12)	(7)	(171)
Oxygen	–	100	88	89	60	48	19	34	41
Number of articles	(0)	(3)	(8)	(9)	(20)	(23)	(43)	(71)	(177)

Note further from Table 4.8 that the proportion of oxygen articles which was theoretical was extremely high at first (as predicted) and declined sharply after 1788. We attribute this drop primarily to the fact that the new paradigm had become successful and was being used to guide "normal" research.[14] There was less need for theoretical battle with the defenders of phlogiston. More support for this hypothesis is provided by a breakdown of the "oxygen" category into its three components across all time periods: the articles classified "antiphlogiston," which were by definition directly involved in the paradigm battle, were 67% (16 of 24) theoretical; those classified "oxygen" (that is, those which actually used the concept oxygen) were 56% (53 of 94) theoretical; while those classified "new nomenclature" and, therefore, least involved in the dispute, were only 8% (5 of 64) theoretical.[15]

On the other hand, the British articles seem to disconfirm this hypothesis, although the small numbers (only five "oxygen" articles) make any conclusions

[14]Again, part of the decline may also be due to the effects of the French Revolution on publications, but we have no evidence to indicate that this was the case.

[15]These results are based on the dichotomous scale of theoretical level. Using the three category scale the results are as follows: "antiphlogiston": 17% "descriptive," 33% "mixed," and 50% "theoretical"; "oxygen": 20% "descriptive," 48% "mixed," and 32% "theoretical"; "new nomenclature": 66% "descriptive," 31% "mixed," and only 3% "theoretical."

TABLE 4.9
Percentage Distribution of Oxygen Articles over Time
by Level of Quantification, Controlling for Country

	1760 -71	1772 -77	1778 -80	1781 -84	1785 -86	1787 -88	1789 -90	1791 -95
				France				
Qualitative	0	0	6	6	35	64	77	90
Number of articles	(18)[a]	(28)	(17)	(32)	(23)[b]	(14)	(39)	(42)
Quantitative	0	17	50	47	60	87	81	92
Number of articles	(3)	(18)	(14)	(15)	(20)	(16)	(16)	(36)
				Britain				
Qualitative	0	0	0	0	0	0	20	33
Number of articles	(0)	(6)	(1)	(5)	(2)	(8)	(10)[a]	(6)[a]
Quantitative	0	0	0	0	0	0	0	50
Number of articles	(1)	(0)	(0)	(5)	(5)	(2)	(5)	(2)[a]

[a] Includes one neutral article.

[b] Includes two neutral articles.

tenuous. Even so, by 1791-95 most theoretical articles (three of four) used the oxygen paradigm.

Thus both parts of Hypothesis (H2) are supported by the French data: the proportion of theoretical articles increased until 1785-86, then decreased, and the proportion of theoretical articles among oxygen papers was high at first and then declined. Hypotheses (H6) and (H12) are also supported: the proportion of papers using the oxygen paradigm was higher among theoretical papers than among nontheoretical; theoretical papers shifted to the oxygen paradigm before nontheoretical. The British data support only the first part of Hypothesis (H2).

Table 4.9 shows clearly that, as predicted, French quantitative articles shifted to oxygen well before qualitative ones: the proportion oxygen among quantitative articles was 17, 50, and 47 in 1772-77, 1778-80, and 1781-84, respectively, while among qualitative articles it was only 0, 6, and 6.[16] Furthermore, the proportion oxygen was higher among quantitative articles in every period. Note that the differences are greater than in the case of theoretical level, reflecting the avoidance of quantitative argument by phlogiston chemists: as one might expect,

[16] Again, eliminating Lavoisier's articles from the table does not change the conclusion. (See Appendix C, Table C.7.)

TABLE 4.10
Percentage of French Articles which Were Quantitative, by Paradigm

	1760 -71	1772 -77	1778 -80	1781 -84	1785 -86	1787 -88	1789 -90	1791 -95	Total
Phlogiston	15	35	30	21	38	29	25	43	29
Number of articles	(20)	(43)	(23)	(38)	(21)	(7)	(12)	(7)	(171)
Oxygen	–	100	88	78	60	61	30	47	50
Number of articles	(0)	(3)	(8)	(9)	(20)	(23)	(43)	(71)	(177)

the defenders of the old paradigm reacted in theoretical, but not quantitative, terms.

Again it is useful to look at the percentages within paradigm. It is obvious from Table 4.10 that "oxygen" articles were more likely to be quantitative in every time period than were "phlogiston" ones. As predicted there was also a clear decline in the proportion of phlogiston articles that were quantitative once the new paradigm took the offensive (after 1785), as predicted. The percentage in 1791-95 is misleading because there were very few phlogiston articles at all in this period.

The proportion of oxygen articles that were quantitative also declined. This is due to the extremely high initial level of quantitativeness, but the decline was greater than we expected, especially in 1789-90. There are two possible reasons for this: (1) that there were new problems facing the oxygen chemists, not amenable to quantitative methods; and (2) that quantification (of the sort we are measuring) had not yet been institutionalized. We will assume the first possibility is correct (Chapter 1, footnote 20, p. 17).

The British data again tend to disconfirm the theory, although the numbers are, as before, too small to allow solid conclusions. We suspect that, although a few nontheoretical and qualitative articles had switched before the first theoretical or quantitative ones, these relationships would look different in subsequent years.

The final shift hypothesis concerns age. For the purpose of testing it, we have dichotomized the age variable at forty and compared those papers by authors aged forty or younger with those by authors aged forty-one or older. In this particular case, the articles of Lavoisier "interfered" with the results,[17] so they were eliminated from Table 4.11.

[17]Since Lavoisier was born in 1743, his articles fall in the "young" category until 1784. Since they dominate the "oxygen" category, the results make is appear that young authors began to switch and then were overtaken by older authors (i.e., Lavoisier). (See Appendix C, Table C.9.)

TABLE 4.11
Percentage Distribution of Oxygen Articles over Time by
Age, Controlling for Country

	1760 -71	1772 -77	1778 -80	1781 -84	1785 -86	1787 -88	1789 -90	1791 -95
				France				
Forty or younger	0	0	0	10	35	100	86	100
Number of articles	$(6)^a$	(14)	(9)	(20)	$(20)^a$	(16)	(37)	(59)
Forty-one or older	0	7	0	0	0	30	53	59
Number of articles	(13)	(28)	(14)	(20)	$(10)^a$	(10)	(15)	(17)
				Britain				
Forty or younger	0	0	0	0	0	0	0	33
Number of articles	(0)	(2)	(0)	(0)	(0)	(1)	(1)	$(3)^b$
Forty-one or older	0	0	0	0	0	0	14	40
Number of articles	(1)	(4)	(1)	(10)	(7)	(9)	$(14)^a$	(5)

[a]Includes one neutral article.

[b]Includes two neutral articles.

With the exception of Bayen's two articles in 1772-77, the French data support the hypothesis that younger authors will switch first: the proportions of articles that used oxygen among the papers written by men aged forty or younger were 10, 35, and 100 in 1781-84, 1785-86, 1787-88, respectively, while for authors over forty the respective proportions were 0, 0, 30. Furthermore, the fact that almost all of the papers by young authors supported oxygen after 1786 makes it quite clear that in 1791 Lavoisier was correct in thinking that the younger generation had accepted the new paradigm (Chapter 2, p. 37).

The results for the British papers appear to be just the opposite — the papers by younger authors switch later (although none of the three in 1791-95 supported phlogiston). However, this is partly due to the fact that the British authors were so old. If we dichotomize at the median age (52), we find that in 1789-90, 14% (2 of 14) articles by those over 52 supported oxygen while none (of 1) by those under 52 did; but in 1791-95, 50% (3 of 6) by those under 52 supported oxygen (another 2 are neutral) while none (of 2) of those by older men did.

Taken collectively, Tables 4.6-4.11 show that the earliest converts were French, that they tended to write theoretical and/or quantitative articles, and that they were young. Leaving British authors aside, since their resistance is well known and has been noted in several places, those resistant to change were older and wrote qualitative and/or nontheoretical articles. These are, of course, generalizations, and there are exceptions. When we look at all articles supporting phlogiston after 1786, of which there were 26, we find them divided among authors as follows: Sage — 15; Delamétherie — 5; Pelletier — 2; Dize — 2; Monnet — 1; and Proust — 1. Thus, two men were responsible for 77% (20 of 26) of all the articles in support of phlogiston after 1786. Furthermore, of the seven articles still supporting phlogiston after 1790, Sage wrote five and Delamétherie the other two.[18]

Looking at the finer breakdowns, we find that Monnet (1), Delamétherie (4), and Sage (2) wrote all seven of the theoretical articles and that Delamétherie (2) and Sage (4) wrote six of the eight quantitative articles supporting phlogiston after 1786. As far as age is concerned, only five articles classified "phlogiston" were written by men under forty after 1786: two by Pelletier, two by Dize, and one by Proust. All five articles were nontheoretical, and only two (by Dize and Proust) were quantitative.[19] All three men were apothecaries: Proust studied with H. M. Rouelle (younger brother of the famous Rouelle we discussed in Chapter 2), and both Pelletier and Dize were assistants to Darcet. Thus the delay in their acceptance of the new paradigm (nomenclature) may have been due to their teachers' influence. The remaining three were all over forty: Monnet was born in 1734, Sage in 1740, and Delamétherie in 1743. As well as being rather old — 53 at the time (1787) his article was published — Monnet had been a long-time enemy of Lavoisier (Rappaport, 1969), so his resistance to the new paradigm is readily comprehensible, even overdetermined. Both Sage and Delamétherie were peripheral to chemistry, Sage being an eminent mineralogist, one of the fields for which the phlogiston theory had been created. Therefore, Sage was not only too old, perhaps, to change easily (although there were converts who were even older, such as Baumé), but he was in a field for which phlogiston theory was especially useful, and he was undoubtedly less conscious of the anomalies upon which the oxygen chemists focused their attack. Delamétherie was

[18]There were four additional pro-phlogiston articles after 1786 by noncited authors: two in 1789-90 by Couret and Coze and two in 1791 by Lechandelier (who also published in 1760 and was, therefore, quite old by the time of the dispute) and Fougeroux (whose article was presented to the Académie in 1788 and was classified on the basis of the old nomenclature), who was not primarily a chemist, and who was also an older man (56).

[19]Furthermore, all five articles were classified "phlogiston" on the basis of the old nomenclature. Since these men were apothecaries, not chemists, it is likely that the possible advantages of the new nomenclature were not apparent to them even if they had been favorably inclined toward the new theory. Since the articles are the basis of the analysis, we classify them phlogiston, but we note that nomenclature, while part of the paradigm, may not be essential for acceptance of other parts of the paradigm by men peripheral to the field. As noted in Appendix A (p. 130), we found other evidence of this in the case of some British articles. Since this issue raises questions about the "gestalt" nature of the paradigm, we shall consider it further in Chapter 7.

a "natural historian" and editor of the *Observations sur la Physique,* as we have seen. There is some question as to whether the term "scientist" should apply to him at all, but clearly he was not directly involved in chemistry and was also over forty.

Thus, the types of men who converted late or not at all add further support to our theory. They were peripheral to the field and thus able to avoid the anomalies longer; they wrote primarily nontheoretical, qualitative articles; and they were well along in their lives and careers when the crisis became acute.[19]

We have seen, then, that the French papers supported the theory and the hypotheses concerning conversions: French articles switched to support of the oxygen paradigm before the British (H11), theoretical papers switched before nontheoretical ones (H12), quantitative papers switched before qualitative ones (H13), and younger authors switched before older ones (H14). The British data generally appear to contradict the hypotheses, although the small numbers are subject to wide fluctuations. However, in each case, although the initial support for oxygen came from the "wrong" group (nontheoretical, qualitative, old), by 1791-95 the proportion supporting oxygen was higher among theoretical, quantitative papers and those by younger authors.

There was also support from both countries for the hypotheses that there would be an increase in theoretical and quantitative articles as the dispute developed, with subsequent declines in the level of theory. Quantitative gains, however, were neither as substantial nor permanent as expected, and this was attributed to new problems for the new paradigm. The Revolution also appeared to draw out theoretical opinion: the proportion of articles not indicating a stand declined, and there were few neutral papers. Finally, we get a clear picture of an attack on phlogiston in theoretical, quantitative terms, a strong initial response by phlogiston chemists, and the rapid triumph of the oxygen theory in France, with resistance almost to the end of the period in Britain.

---

[19]In the final chapter I will discuss some of the implications of the characteristics of resistors for the status of phlogiston as a paradigm in the 1760s.

5
Models of Revolution

In order to do a causal analysis, we have taken an "article" to be the basic unit of analysis and restricted the articles to those which deal with the relevant paradigms. Therefore, as with the data for the shift hypotheses, all articles classified as "no information" have been excluded from the following analyses.[1] Further, because age is an important variable, all articles by authors for whom we were unable to find a birthdate were eliminated. Thus, the analysis has been carried out on the basis of the "analytic" sample of 409 articles, 351 French and 58 British.[2] Although this means that we do not have a universe of all articles from 1760-1795 in the following analyses, we still have the universe of all articles which stated a paradigm choice and which were written by authors important enough to be noted by historians.

[1]The following models were also examined with data including these articles in two ways: (1) by putting all the "no information" articles into the "neutral" category. As might be expected, in view of the strong association of this category with "descriptive" articles, as shown in Appendix A, this strategy depressed coefficients involving the level of theory, but this was not a very serious effect, nor were any of the other relationships affected significantly except age and time in France and effects on "oxygen" in Britain. (See Table C.13) (2) The "no information" articles were distributed into the three categories of "phlogiston," "neutral," and "oxygen" according to our closest approximation based on information about the authors from secondary sources and from their other articles. The coefficients in most cases were increased (in absolute value), but the inherent possibilities of error and bias outweighed the advantages of "better" results, so we chose to use the "analytic" sample results. (See Table C.14.)

With these "no information" articles included the relationship between age and time in France became positive, indicating that the less engaged authors grew older in contrast to those involved in the revolution.

[2]See Appendix A for a discussion of the various samples. There were no noteworthy changes in the relationships among the variables relative to the samples used in preceding chapters. (See Tables C.15-C.17.)

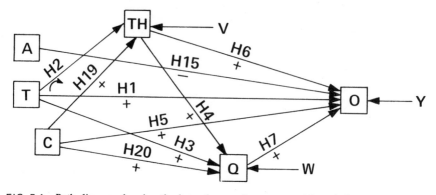

FIG. 5.1   Path diagram showing the hypotheses, where: + = positive relation, − = negative, ⌐ = curvilinear, A = age of author; O = phlogiston/oxygen; T = date minus 1760 (so that time (0) is 1760); C = country, with France counted as "high"; TH = level of theory; and Q = level of quantification (quantitative is "high"). The arrows from the literal variables (lowercase letters – v, w, y) are the residual paths, representing the effects of all variables not in the system (including random effects and measurement errors).

Causal modeling is a method of data analysis which attempts to explain empirical findings in a manner that reflects the total system or structure of relationships which underlies the phenomena under study [3]:

> When models are used in theory building and systems analysis, the goal is to define a set of equations which, in some sense, corresponds to actual causal processes in the real world. . . Defining a causal structure of this kind requires specifying the network causal paths that exist between variables, and identifying the parameters of causation so that one knows how much each variable affects another. [Heise, 1969, p. 41]

The technique of ascertaining the causal links is the calculation of path coefficients (Wright, 1921, 1934, 1954).

Before turning to detailed models of the theory which take into account several predictable interactions and nonlinearities, we begin with a basic model representing the average effects of the variables over entire period. Figure 5.1 shows this basic path analytic representation of the theory.

With this model representing the theory, an examination of the (standardized) path coefficients will show if this version is compatible with the more elementary version of Chapter 4 (that is, if the more powerful assumptions necessary for path analysis result in obvious discrepancies) and additionally they will provide us with estimates of the relative effects of the variables when the others are held constant. Figure 5.2 depicts the results.

The results parallel those found in Chapter 4. Taking the dependent variables in order, we see that the linear effect of time on the level of theory was very slight (−.04). This is to be expected since we have predicted a curvilinear relation-

---

[3] The notion of causality is obviously complex. For a general discussion of the idea as it is used in connection with causal modeling, see Blalock (1964, Chapter 1).

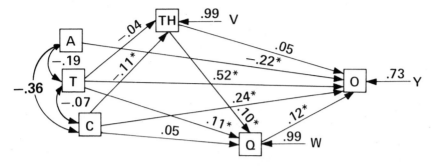

FIG. 5.2   Path diagram showing path coefficients. * = $p < .01$.

ship and this model assumes a linear one. Country had a weak *negative* effect (−.11) on the level of theory: the British were more theoretical, contrary to Hypothesis (H19). As for quantification, the effects of both time (.11) and theory (.10) were weakly positive while country (France) had a negligible (.05) positive effect. Not very much of the variance in theory or quantity is explained.

Turning to our primary concern − the support of the paradigms − the results are much better. Time was the strongest variable (.52), accounting for more than 25% of the variance directly, an expected influence since a paradigm shift is, above all, a temporal process. The effect of country (France) and age were about equal (.24 and −.22, respectively), supporting (H5) and (H15). The moderately strong correlation between age and country (−.36), while not part of the theory, was to be expected from the results of Chapter 4: British authors tended to be older than their French counterparts. Furthermore, the indirect effect of country on use of oxygen through age was .08 (−.36 × −.22), adding another third to the direct effect. Finally, quantification had a slight effect, while that of theoretical level was negligible, thus providing only weak support for (H6) and (H7), our hypotheses that quantitative and theoretical papers would disproportionately use oxygen.

In addition, however, we see that age had a negative correlation with time (−.19), supporting the argument in Chapter 1 that a revolution attracts new, younger authors (potential gains outweigh risks). I will consider this later in connection with French-British differences.

This average model, then, provides some support for the theory. However, some relationships require further elaboration. We have hypothesized that the relationship of theoretical level and time was curvilinear, and the tabulations in Chapter 4 showed a similar tendency for the relationship between time and quantification. We also expect the relationship between theoretical level and support for oxygen to change over time as phlogiston theorists reacted and then withdrew, and likewise, the relation between quantification and oxygen to some extent. In addition, we discovered interaction between some of these relationships and country: the relationships among level of theory, quantification, and age to some extent were reversed in Britain as compared to France. To depict

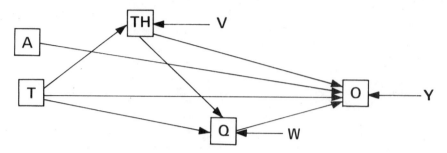

FIG. 5.3   Path diagram of model run within each country.

these relations, it is necessary to look at slightly modified models for each country and for different time periods: 1760-77 (normal science to anomaly), 1778-84 (crisis), 1785-88 (debate and conversion), 1789-95 (consolidation, new paradigm period). A diagram of an appropriate model is presented in Fig. 5.3. For conciseness the coefficients are given in Tables 5.1 and 5.2.

In the analyses to follow, we will first discuss the unstandardized coefficients of a given model, comparing across groups, and follow this with a discussion of standardized (path) coefficients, comparing within groups.[4] In Table 5.1, keep in mind that the coefficients of different variables cannot be compared with each other (e.g., within time periods) since they are based on different units of measurement.

For French articles, the overall results support Hypotheses (H3), (H4), (H1), (H6), (H7), and (H15) almost completely: the increase in time led to an increase in quantitativeness (.01); the increase in theoretical level led to an increase in quantification (.02); the probability that an article would use oxygen increased over time (.08); the more theoretical a paper was, the more likely it used oxygen (.03); quantitative papers were more likely to use oxygen than qualitative ones (.30); and younger authors were more likely to write articles using oxygen (−.02).

The British articles show some interesting contrasts. First, the effect of time on quantification was much less (.002 versus .01) than in France, primarily because British chemists were predominantly phlogistonists, and phlogistonists tended to avoid quantitative work. Second, the effects of theory and quantity on phlogiston/oxygen are opposite from those in the French data (−.02 and −.06 vs. .03 and .30), although the coefficients in the British data are close to zero. Noting that all the variation in the dependent variable occurred in the final period (1789-95), we see that this reflects the data of chapter 4: theoretical papers tended to remain phlogistonist, as did the few quantitative ones (Tables 4.7 and 4.9). If we had extended the time period of British publications for a few more years, these relationships probably would have changed. Finally, the effect of age on phlogiston/oxygen was greater in the British articles (−.03

---

[4]For a discussion of the differences and the statistical implications thereof, see Blalock (1971), Cain and Watts (1970), Schoenberg (1972), and Wright (1960).

TABLE 5.1
Unstandardized Coefficients

Independent variables	Dependent variables				
	1760-77	1778-84	1785-88	1789-95	Overall
	France				
	Level of theory				
Time	$.08^a$	$.22^b$	−.25	−.16	−.01
	Level of quantification				
Time	$.03^b$	−.01	.001	$.08^b$	$.01^b$
Level of theory	*.05*	.04	−.005	−.001	*.02*
	Phlogiston/oxygen				
Time	−.004	−.02	$.26^b$	.04	$.08^b$
Level of theory	$.07^b$	$.14^b$	.06	.008	.03
Level of quantification	*.18*	$.72^b$	$.48^b$	.08	$.30^b$
Age	.000	−.006	−.02	$−.03^b$	$−.02^b$
Number of cases	67	78	73	133	351
	Britain				
	Level of theory				
Time	1.14^b	1.02^b	$−.98^b$	*−.56*	−.05
	Level of quantification				
Time	*−.26*	−.02	−.14	−.006	.002
Level of theory	*.11*	.03	.05	−.001	.03
	Phlogiston/oxygen				
Time	−	−	−	.17	$.05^b$
Level of theory	−	−	−	−.05	−.02
Level of quantification	−	−	−	−.15	−.06
Age	−	−	−	−.03	$−.03^b$
Number of cases	7	11	17	23	58

[a]Italicized numerals: $= p < .05$.

[b]$p < .01$. Values should be compared *across*, not within, groups.

versus −.02). While this is intriguing, it is not statistically significant (see Table 4.11).

Dividing the data into time periods provides additional details and specifies the hypotheses in useful ways. We will look first at each dependent variable in turn and then at the model as a whole.

Among French articles, the level of theory increased during the early part of the Revolution, the crisis period, until 1785-88, as shown by the two positive

coefficients (.08, .22). It then decreased, as shown by the subsequent two negative coefficients (−.25, −.16), until the end of the period under study. This
supports (H2). Both the increase and the decrease were sharpest in the middle
periods (.22 in 1778-84 and −.25 in 1785-88), those surrounding the peak of the
debate. The result is that there was no overall linear relationship between time
and level of theory, but the sequence of regressions within time periods supports
the predicted curvilinear relationship strongly. For British articles, the results
are similar (the magnitude of the coefficients can be ignored because the subsample sizes are so small, but the pattern is quite clear): an increasing amount of
theoretical work followed by a decline.

The picture is different for level of quantification: among French articles
the proportion of quantitative articles increased during the first period (.03),
then decreased slightly (−.01), remained unchanged in 1785-88, and then increased sharply (.08) in the final period. Since level of quantification is a dichotomous variable, the regression coefficient can be expressed as a change in the
probability of finding an article of a certain type. Thus, in the first seventeen
years the probability that an article would be quantitative was increased by 51%
(.03 X 17); over the next six it declined by 6%; and it remained unchanged until
the last six years, when it increased again by 48% (.08 X 6), if the level of theory
is held constant. The overall effect, however, was only 1% per year.[5] Further,
the hypothesis that the more theoretical an article is the more likely that it will
be quantitative (H4) was supported only in the first two periods.

The explanation for this probably lies partly in the reaction of the phlogiston
chemists, noted in Chapter 2 (p. 31 and footnote 19) and discussed in relation
to Tables 4.7 and 4.9: the phlogistonists increased their output of theoretical
articles for some time while to a large extent avoiding quantitative work. Since
in this case the new paradigm was superior in its ability to account for quantitative results, and since this ability, and a concomitant emphasis on it, was instrumental in the overthrow of the old paradigm, it is not surprising to find phlogiston
chemists avoiding quantitative arguments especially in their theoretical papers.

---

[5]That the overall effect − in this case a predicted increase over the 36 years of 36% in
the probability that an article will be quantitative − is not equal to the sum of the piecemeal
effects may seem surprising. However, it is a statistical artifact due, in a sense, to where the
cut-points happen to fall. The influence of time on quantitativeness in Britain − all negative
segments, but an overall positive effect − may seem even more anomalous at first glance.
However, if one plots the lines, using both the slopes and the intercepts (and keeping in
mind that we are controlling for the effect of another variable), the picture looks somewhat
like the following:

Such a reaction, of course, depresses an otherwise positive effect of both time and theory on quantification.

Two kinds of interaction might be examined in order to see if this is a reasonable explanation: first, an interaction between level of theory and quantification. If an article were either nontheoretical or qualitative (or both), it could avoid the sharpest (quantitative) anomalies, whereas if it were both theoretical and quantitative, it would be extremely difficult to avoid the impact of the anomalies. This is treated later in this chapter (pp. 96-98). The second sort of interaction involves treating phlogiston/oxygen as a control variable and studying the relationship between level of theory and quantification. This was done, but the results were not as expected. There was, indeed, a negative effect of theoretical level on quantification among the phlogiston papers; however, the negative relationship was even stronger among the oxygen papers. Examination of the underlying distribution provides an explanation for this result. The actual relationship appears to have been curvilinear throughout much of the time involved: both descriptive and highly theoretical articles had extremely high proportions of quantitative articles, while articles that fell in the middle of the theoretical scale had a lower proportion.[6] Since the number of theoretical articles was relatively small, they had little effect on the calculation of the coefficient. The result therefore is a negative slope (descriptive articles were highly quantitative, mixed were low). Furthermore, since the earliest oxygen articles were highly theoretical and wholly quantitative, the proportion quantitative could only remain constant or decline. In any case, among these oxygen papers *highly* theoretical articles were always at least as quantitative as those less theoretical.[7]

For British articles, the results are mixed. The effect of time on quantification has already been discussed (see footnote 5, p. 88). The effect of theoretical level, on the other hand, supports (H4): there was a slight positive effect on quantification (though it was not statistically significant), except for the last period (1789-95), when the relationship was essentially negligible. Given that all the articles until this last period used phlogiston, however, the explanation advanced for French articles does not hold for the British. For British articles, the positive effect of theoretical level on quantification implies either that quantitative anomalies had little effect or that they were not perceived.

As for choice of paradigm, the results for French articles support the theory, and the breakdown by time intervals makes the sequence of events clear. First,

[6]This suggests that a further refinement would be to include a squared term for theoretical level.

[7]It is quite possible that the high proportion of quantitative papers among the least theoretical articles is due to "pseudoquantification" (recipes), the measurement problem discussed previously and in Appendix A. This should be more of a problem for purely descriptive articles, because it is easier to decide if a paper is really quantitative when the results are being explained, as in the more theoretical articles.

we note that time, which had a strong overall effect and which may seem to be a trivial and obvious variable at first glance, had very little direct effect except in the period 1785-88. This is credible if we remember that only Lavoisier (and his coauthors, Laplace and Meusnier) and Bayen questioned phlogiston before 1785. Thus, as the number of authors increased, there was even a slightly negative direct effect of time on the use of the oxygen paradigm. During 1789-95, on the other hand, most of the authors who were going to switch but had not already done so switched early (1789-90) in the period (see Table 4.7 or 4.9), and there was no room for further change.[8]

Second, in the two early periods, the more theoretical an article was, the more likely it was to be antiphlogiston (pro-oxygen). This relationship weakened by 1785-88 and was almost absent by 1789-95 due to the reaction of the phlogiston chemists, which was pointed out in the discussion of Table 4.7 above. Two additional factors contributed to the lack of relationship after 1788: (1) that there was relatively little room left for change and the few resisters still argued in theoretical terms; and (2) that most of the oxygen chemists were engaged in normal science under the new paradigm, and the theoretical content of their articles was lower than in earlier periods (see Table 4.8).

For all periods, quantitative articles were more likely to be pro-oxygen or antiphlogiston than qualitative ones. This difference was especially strong during the two middle periods when the dispute was raging, but it weakened greatly in 1789-95. Again, we attribute the weaker effect in the last period to the lack of authors left to switch after 1789 and to the few defenders left arguing to some extent in quantitative terms. Furthermore, by this time there was a tendency for even those authors who did primarily qualitative and/or nontheoretical work to join the bandwagon, as shown in Tables 4.7 and 4.9, and to a great extent this nullified the predicted relationship.

Finally, age had almost no effect until after 1784, for it was not until then that we find converts other than Lavoisier and Bayen. That is, when the new paradigm did begin to win converts, they were indeed young (shown also in Table 4.11). However, it is probable that another as yet unmeasured variable kept age from having an even stronger influence in 1785-88: social proximity to Lavoisier. As noted in Chapter 2, conversion depended to a great extent on personal contact with Lavoisier (or, later, with other members of the oxygen group). Thus, the first converts were from the group in Paris which was very close to Lavoisier, and were often his collaborators. These men were not old, but neither were they very young – many of them were the leaders of the field.

Furthermore, Lavoisier and Berthollet alone accounted for almost half the articles written during this period. Lavoisier was then over forty and Berthollet was just under forty. Lavoisier's articles, then, had a particularly depressing effect on this coefficient. If they are eliminated from the regression (as they were from

---

[8]We must also bear in mind that part of this lack of effect of time is due to the fact that we are partially "controlling for" the effect of time (by running the model within time periods) and, thus, constraining its effect "artificially."

Table 4.11), then the coefficient changes from −.02 to −.05, a substantial difference. [9]

Once this initial, middle-aged group converted, however, age had a strong negative effect, until the time that all the youth and most of the older men had also converted. In the last period, a difference of 20 years (e.g., between a neophyte of 30 and an established scientist of 50) resulted in an increased probability of 30% of using the oxygen paradigm, even though all but a few had already shifted.

The process as a whole is now more clear. Time affected level of theory in a curvilinear fashion and had its greatest effect in the middle two periods, when the crisis and debate were most intense. Time had its greatest effect on quantification during the first and last periods, while theory had its greatest effect on quantification in the first two periods, when the anomalies became apparent and the crisis was developing. As far as choice of paradigm is concerned, time had its greatest effect in the period when major conversions were taking place, whereas level of theory and quantification had their greatest effect earlier, when the crisis was building. Finally, age most greatly affected paradigm choice during the periods of major conversion and consolidation, indicating that the defenders of phlogiston were older men.

The examination of unstandardized coefficients has shown that the theory is generally supported: most of the hypotheses are confirmed, and the anomalies are due to the reactions of the "counterrevolutionaries" and the problems of disentangling these effects. We will now look at the path (standardized) coefficients in order to compare the effects of the variables to one another within each time period, and to examine selected indirect effects. The data are given in Table 5.2.

We begin the discussion with the French articles. The period 1760-77 represents a period primarily of "normal" science, when anomalies became apparent to few chemists, notably Lavoisier and Bayen. Time had a moderately weak effect on theoretical level (.16), accounting for only 3% of its variance, while

---

[9]Lavoisier's articles also depress all other coefficients in this period, except that for quantification on paradigm choice. Omitting his articles from the regression gives the following results for effects on paradigm choice: time − .54*, theoretical level − .10*, and quantitativeness − .06 (* = $p < .01$). On this basis, then, the model fits the theory even better than before. And this is as it should be. Knowing that Lavoisier had converted long before 1785, one would expect that time would have no effect on his paradigm choice in the period under consideration. As far as theoretical level is concerned, we can reasonably assume that some of his articles were devoted to working out details and applications of the new paradigm, rather than all of them being devoted to theoretical dispute − especially since he had supporters during this period. As for quantification, we can now only conclude that aside from Lavoisier himself, it had relatively little effect on paradigm choice at this time.

While Lavoisier's articles, then, depressed most of the coefficients in this period, they had little effect in other periods except to slightly depress the effect of time on all variables and slightly enhance the effects of theoretical level and quantification in the first two intervals.

TABLE 5.2
Path Coefficients[a]

Independent variables	Dependent variables				
	1760-77	1778-84	1785-88	1789-95	Overall
	France				
	Level of theory				
Time	.16	.23	−.13	−.10	−.03
(Pvt)	.98	.97	.99	.99	1.00
(R-squared)	.03	.05	.02	.01	.001
	Level of quantification				
Time	.26	−.02	.002	.27	.12
Level of theory	.22	.19	−.02	−.01	.09
(Pwg)	.93	.98	1.00	.96	.99
(R-squared)	.14	.03	.00	.07	.02
	Phlogiston/oxygen				
Time	−.04	−.06	.32	.09	.55
Level of theory	.36	.38	.13	.03	.07
Level of quantification	.20	.43	.25	.04	.15
Age	.01	−.07	−.14	−.37	−.18
(Pyo)	.89	.77	.90	.92	.75
(R-squared)	.20	.40	.19	.15	.44
	Correlation of age and time				
	−.02	−.18	.05	−.01	−.28
Number of cases	67	78	73	133	351
	Britain				
	Level of theory				
Time	.81	.69	−.52	−.33	−.10
(Pvt)	.59	.72	.85	.94	.99
(R-squared)	.65	.48	.27	.11	.01
	Level of quantification				
Time	−1.30	−.05	−.39	−.02	.03
Level of theory	.73	.13	.26	.002	.17
(Pwg)	.59	.99	.82	1.00	.98
(R-squared)	.65	.01	.32	.00	.03
	Phlogiston/oxygen				
Time	—	—	—	.26	.50
Level of theory	—	—	—	−.13	−.07
Level of Quantification	—	—	—	−.09	−.05
Age	—	—	—	−.35	−.46
(Pyo)	—	—	—	.84	.59
(R-squared)	nv[b]	nv	nv	.30	.35

(continued)

TABLE 5.2 *(continued)*

Independent variables	Dependent variables				
	1760-77	1778-84	1785-88	1789-95	Overall
	Correlation of age and time				
	.20	.59	.12	−.30	.31
Number of cases	7	11	17	23	58

[a] Values in the table should be compared *within,* not across, time periods.

[b] nv = no variance. Pij are the paths from the residuals to the relevant variables and represent the effect of all causes not in the model on the dependent variable. They are equal to the square root of 1 minus R-squared.

both time (.26) and level of theory (.22) had moderate effects on quantification, accounting for 14% of its variance.

As for paradigm choice, level of theory had the greatest effect (.36) and level of quantification somewhat less (.20). Interestingly enough, the direct effect of time, when theory and quantification are controlled, was very weakly negative (−.04) as noted in the discussion of unstandardized coefficients. However, the indirect effects of time through theory, .06 (.16 X .36), and quantification, .05 (.26 X .20), were positive and together outweigh the direct effect (.11 versus −.04). Thus, we see that, given anomalies, the passage of time will lead directly to an increase in the typical level of theoretical and quantitative work. The increasing proportion of theoretical and quantitative articles in turn makes the anomalies more severe and results in a tendency to reject the given paradigm. (The indirect effect of level of theory through quantification to paradigm choice (.04 = .22 X .20) was negligible.)

During the next period (1778-84), when the old paradigm was coming under increasingly severe attack from Lavoisier, and phlogistic chemists were reacting to this pressure, the picture changes somewhat. The effects of time on theoretical level and theoretical level on quantification were still moderate, but that of time on quantification disappeared. Furthermore, with respect to paradigm choice, the direct effect of quantification was slightly stronger than that of the amount of theory (.43 vs. .38). However, the indirect effect of theory through quantification was .08 (.19 X .43), so its net effect was slightly greater. These strong effects show that at this point in the revolution (a period of crisis), the problems were brought into sharp focus by attempts to account for them theoretically and to deal with them quantitatively.[10] Again, the direct effect of time was weakly negative

[10] The magnitude of the coefficients produced a larger explained variance of paradigm choice in this period than in the preceding one (40 versus 20%). Both of these percentages are impressively high considering that our most obvious, and strongest overall, variable − time − has almost no direct effect.

(−.06), but the indirect effect through theoretical level was sufficiently positive, .09 (.23 X .38), to give a net result of essentially no effect.

Up to this point, the data for British articles displayed similar patterns, although the coefficients were unrealistically high (due to small ns). The major difference, which persisted until 1785, was the opposite relationship of age and time, with French authors becoming younger and British becoming older.

The next period (1785-88) represents the major period of conversion. Time had a weak negative effect (−.13) on level of theory.[11] As above, both time and theoretical level had practically no effect on quantification.

Now that the new paradigm was spreading, time had the strongest effect (.32) on paradigm choice, quantification had somewhat less effect (.25), while theory (.13) and age (−.14) were relatively weak. However, as in the case of the unstandardized coefficients, Lavoisier's articles reduced the effect of age. In fact, his articles depressed all of the coefficients in this period, except that of quantification: time clearly did not affect the paradigm choice of his articles, nor did the theoretical level of his articles. If we remove his articles from the regression, the results are as follows: the effect of time is .65, of level of theory is .21, of quantification is .03, and of age is −.35. The resultant R-squared is .49! Although this is not the same sample as that which includes Lavoisier's articles, the respective variances are almost identical and it is clear that our model represents the process of paradigm change better if we ignore Lavoisier's articles in this time period. This result is encouraging, because he had shifted much earlier.[12]

The relatively strong effect of time is even more remarkable in view of its small variance. If we compare these path coefficients with the unstandardized coefficients for the same period, we see the consequence of differences in units of measurement: the unstandardized coefficient of the effect of quantification on paradigm choice was almost twice as large as that of time on paradigm choice, but time had a stronger standardized effect. That is, time explained a greater amount of the variance than did quantification; or, put another way, time was a better predictor of paradigm choice than was level of quantification.

[11]The small value of this coefficient is somewhat surprising in view of the fact that the unstandardized coefficient was stronger during this period than in the preceding one (Table 5.1). This is a result of the smaller standard deviation of time during this period in comparison to the preceding one (1.2 versus 2.3 — see Table C.7) and the relation between path coefficients and unstandard coefficients. It is due to this relationship that it is dangerous to compare standardized coefficients across groups (compare Schoenberg, 1972, and Cain & Watts, 1970).

[12]The values of the coefficients are depressed by authors who switched early ("before their time"), as well as by defenders who were young and/or wrote theoretical and/or quantitative articles. Thus it is to be expected that the elimination of Lavoisier's articles after 1780 strengthens the model. On the other hand, elimination of his articles from the other time periods has little effect, except for reducing the coefficients of theoretical level and quantification on paradigm choice slightly in all periods.

Finally, in the period of consolidation (1789-95), which can also be viewed as the beginning of a new period of normal science, the strongest effect was that of age on paradigm choice (−.37). This accounted for almost all of the explained variance. As in the discussion of unstandardized coefficients, we interpret this to mean that other variables had had their effect, that there were few chemists left to switch, and that the defenders were older men. The only other noteworthy relation is that of time on quantification (.27), although time also had weak effects on level of theory (−.10) and paradigm choice (.09).

For British articles, the results are similar, except for the strong effect of theoretical level on quantification in 1785-88 (.26) and opposite effects of theoretical level (−.13) and level of quantification (−.09) on paradigm choice (1789-95), echoing the findings in Chapter 4.

Another finding of interest is that the average age of the French authors declined over time. Because of the way that this variable has been measured — each author is, in effect, weighted by the number of articles he contributed — it could be due either to more young authors or to an increase in the productivity of young authors and/or a decrease in the productivity of older authors.

Table 5.3 shows that the productivity of older authors was higher than that of younger ones through 1786 (except for 1778-80 when Lavoisier was frequently publishing), after which the productivity of younger authors was higher. On the other hand, the number of younger authors was greater after 1780 (except for 1778-88), supporting the position that the declining average age was primarily due to young men being attracted to the field of chemistry in France, and only secondarily due, in later years, to their greater productivity. The relatively greater decline in productivity of older men in the period 1791-95 (down to .42 as opposed to .95 for the younger men) probably reflects their greater involvement in the declining fortunes of the Académie.

TABLE 5.3
Productivity of French Authors
over Time by Age (Analytic Sample)

	1760 -77	1778 -80	1781 -84	1785 -86	1787 -88	1789 -90	1791 -95
				Forty or younger			
Articles	26	17	21	20	16	37	59
Authors	11	5	11	7	6	11	12
Articles/auth/yr	.13	1.1	.48	1.4	1.3	1.6	.95
				Over forty			
Articles	41	14	26	23	14	18	19
Authors	18	6	9	5	10	7	9
Articles/auth/yr	.13	.78	.72	2.3	.70	1.4	.42

On the other hand, we see that until the very end of our period the average age of British authors increased, in keeping with our discussion of the poorer motivational structure of British science. There may also have been another factor at work: discouragement during the period of crisis as the French chemists solved problems that stymied the phlogistonists of Britain.

In any case, the combination of the strong negative association of age with acceptance of the new paradigm, and the increasingly older age of the British chemical community, suggests that age alone was a major factor in British resistance to the new paradigm, a fact unnoticed by historians — and one which also supports the theory.

FURTHER CONSIDERATIONS

Dividing the data into time periods essentially means examining the effects of the interaction of time with other variables. As noted in the discussion of Table 5.1, and as can be expected from the discussion of the relationship between theory and quantity (Chapter 1, p. 17) leading to Hypothesis (H4), we expect to find interaction in associations among oxygen, theoretical level, and level of quantification. Theoretical articles that are qualitative may well be able to ignore anomalies, but it is particularly difficult for quantitative theoretical papers to do so.

Dichotomizing the control variables and omitting articles from the period 1760 to 1771 (since they precede the revolutionary process), we arrive at Table 5.4.

We will first consider French articles, comparing theoretical with nontheoretical.[13] Time affected the phlogiston and oxygen paradigms in similar ways in both cases and, as in the overall model above, it was the single most important variable. Age had the same effect among theoretical articles as among nontheoretical.

For level of quantification, there is evidence of the strong interaction we expected. If an article was nontheoretical, its quantitative content made little difference, but if it was theoretical, the change from qualitative to quantitative increased by 33% the probability that an article used oxygen. That is, for nontheoretical articles, quantitative measurements made very little difference in choice of paradigm (this may not be true in every time period, however), while for theoretical articles, quantitative techniques were decisive. In essence, quantifying anomalies increased their impact only for authors with theoretical interests, or more precisely, for theoretical articles.

---

[13]We also inspected the same relationships within the time periods used for the major analysis, but the numbers were so small and the new information so minimal that there was no point in including the results.

TABLE 5.4
Regression Coefficients of Oxygen Regressed on Time,
Controlling for Other Variables in the Model

	French		British	
	Nontheoretical	Theoretical	Nontheoretical	Theoretical
Time	$.10^a$ $(.64)^b$	$.09^a$ $(.47)$	$.04^a$ $(.38)$	$.08^a$ $(.62)$
Quantity	$.10$ $(.05)$	$.66^a$ $(.34)$	$-.25$ $(-.19)$	$.07$ $(.06)$
Age	$-.02^a$ $(-.17)$	$-.02^a$ $(-.17)$	$-.01$ $(-.21)$	$-.07^a$ $(-.78)$
(R squared)	.52	.36	.16	.76
Number of cases	203	127	25	32
	Qualitative	Quantitative	Qualitative	Quantitative
Time	$.11^a$ $(.63)$	$.09^a$ $(.54)$	$.05^a$ $(.47)$	$.10^a$ $(.70)$
Theory	$-.01$ $(-.01)$	$.10^a$ $(.23)$	$-.04$ $(-.13)$	$.07^a$ $(.33)$
Age	$-.02^a$ $(-.20)$	$-.01$ $(-.12)$	$-.03^a$ $(-.37)$	$-.06^a$ $(-.77)$
(R-squared)	.51	.33	.28	.81
Number of cases	195	135	38	19

Phlogiston/oxygen

[a] $p < .01$.

[b] ( ) = standardized coefficients.

The standardized coefficients exhibit a similar pattern. Among nontheoretical articles, time had by far the strongest influence on paradigm choice (.64), while the effect of quantification was negligible (.05). Among theoretical articles, on the other hand, while time still had the strongest influence (.47) and age accounted for the same amount of variance as among nontheoretical, level of quantification had a much stronger effect (.34) and was the second strongest variable. If we think in terms of prediction, this indicates that, among nontheoretical articles, we could predict fairly well from the date of an article whether it was likely to use oxygen or not, and that knowing the age of the author would help us somewhat, but knowing whether or not the article was quantitative would not substantially increase our ability to predict.[14] On the other hand, if we were to consider a theoretical article, the date would still be the best predictor, but the knowledge that it was quantitative would be almost as useful. Finally, note that

[14]Keep in mind that these statements concerning the predictive value of each variable assume, as always in multiple regression, that we are holding the other variables in the equation constant.

more of the variance was explained among the nontheoretical articles than among theoretical.

Results for the quantitative-qualitative dichotomy amplify these patterns. First, time had a similar outcome on paradigm choice among both qualitative and quantitative articles (.11 vs. .09). Further, age had double the effect on oxygen use among qualitative articles than among quantitative ones (−.02 vs. −.01). The greatest difference, however, arises from the influence of theoretical level, which was ten times stronger (and in the opposite direction) among quantitative articles than among qualitative ones: .10 vs. −.01. In other words, if an article was qualitative, it made very little difference whether or not it was also theoretical in regard to the likelihood of its using oxygen; whereas if it was a quantitative article, the more theoretical it was, the more likely it was that it used oxygen. Essentially, then, quantitative articles which were nontheoretical could avoid anomalies and remain pro-phlogiston, but those quantitative articles which were theoretical could not ignore them as easily and were more likely to turn to the oxygen paradigm.

Again the path coefficients paint a similar picture. Among qualitative articles time had by far the strongest effect (.63), while age had a moderate effect (−.20) and theory practically none (−.01). Among quantitative articles, however, time still had the greatest effect (.54), while theory had a moderate effect (.23), and age had little effect (−.12). This tells us that, in either case, time was clearly the most important factor, which is consistent with earlier findings and (H1). Where the work was qualitative, whatever anomalies there were did not force the theorist to shift his world-view any more than the nontheorist, although they did tend to influence younger chemists to do so more often than older ones. However, where the work and, therefore, the anomalies were quantitative, their importance for the theorist was greater than were age differences.

Analogous comparisons for British articles are tenuous, since all the variation in the dependent variable took place only in the last six years of the period, and since the numbers are so small. It is safe to conclude, however, that there is evidence of some interaction: the effects of all three independent variables were quite different within the two levels of theory and quantity.

The causal analyses have given us a detailed and distinct picture of the complex and changing relationships among the variables in the model. However, the picture is not a simple one. Using the information regarding the growth of oxygen over time (a logistic curve: flat at the beginning, rising steeply in the middle, and flat again at the end) and the evidence about the interaction of level of theory and quantification, we can consolidate and simplify the system somewhat by relaxing the assumptions of linearity and additivity.[15] As a first step, which is all that we attempt here, we will look at the equation predicting paradigm choice,

[15]For a discussion of these assumptions and the effects of relaxing them, see Johnston (1972, pp. 47-55) and Heise (1969, p. 65). For a discussion of nonadditivity and problems of multicollinearity, see Althauser (1971).

adding nonlinear and multiplicative terms (polynomial terms for time and a term for the interaction of levels of theory and quantification), resulting in the following equations:

French articles —
$$O = 2.02 - .09T + .004T^2 - .00002T^3 - .10Th - .12Q - .02A + .11Z + E$$

British articles —
$$O = 1.87 + .15T - .01T^2 + .0002T^3 - .08Th - .32Q - .03A + .06Z + E$$

where O = oxygen, T = time, Th = level of theory, Q = level of quantification, $Z = Th \times Q$, and A = age.

There are three major features of these equations: the difference between the time variables in each case, due to the lag in Britain; the fact that in each case the other four variables act in the same direction, with theory and the interaction term (Z) stronger in France and quantification and age stronger in Britain; and the fact that the inclusion of the interaction term results in negative coefficients for its components. The resulting R-squared is .49 for France and .41 for Britain, an increase of .06 over the linear, additive model in each case. Thus, not only are these equations a more realistic view of the overall relationships, but they also increase the explained variance 14% for French articles and 17% for British. This would appear to indicate that while the linear approximations are reasonable, more complex relationships should be considered, and this we have done.

OTHER HYPOTHESES

Aside from those concerning recognition, the two remaining hypotheses are that the use of affinities will be associated positively with both level of theory (H16) and quantitativeness (H17). In order to assess these hypotheses, we will look at the relationships during the same time periods used for the models.[16]

Table 5.5 shows that use of affinities had moderate to strong positive correlations with level of theory among French papers (ranging from .28 to .49) and small to large positive correlations among British (ranging from .11 to .49) in all time periods, supporting Hypothesis (H16). The lowest correlations in each country were in 1778-84, the period of Lavoisier's initial attack on phlogiston and the phlogistonists' reaction. During this period the use of affinities was at a low point and other problems were of greater concern, but with the publication

---

[16] Since attitudes toward oxygen were not in question, I used the cited sample to increase the size of the British subsample. The results were identical for the analytic sample.

TABLE 5.5
Correlations of Affinity with Levels of Theory and
Quantification over Time by Country (Cited Sample)
(Pearson's *r*)

	1760-77	1778-84	1785-88	1789-95
		French		
Level of theory	.49	.28	.45	.49
Level of quantification	.05	.07	−.10	.01
Number of cases	(131)	(140)	(115)	(166)
		British		
Level of theory	.26	.11	.49	.26
Level of quantification	−.19	.07	.38	.09
Number of cases	(27)	(17)	(22)	(28)

in 1785 of Bergman's treatise they again became important, especially in the work of Berthollet and Kirwan.[17]

Associations with quantification, however, were negligible, except for some erratic results in Britain, disconfirming Hypothesis (H17), and illustrating the failure of Newtonian inspired attempts to quantify chemistry through the use of elective attractions (cf. Fichman, 1969; Thackray, 1970b). These attempts triumphed momentarily in the law of mass action of Berthollet after 1800.

These findings indicate that while affinities were a useful theoretical tool, they did not typically follow from or lead to quantification, at least of the kind we are concerned with. This last result, while possibly implying that use of affinities was not pressure to quantify, as stated by Guerlac, may also reflect the inability of eighteenth century chemists to quantify chemical attraction in spite of recurring attempts. Furthermore, quantification was measured in terms of weights of reactants and products in this study, and this particular form of quantification does not appear to have been related to affinities until the later work of Berthollet, around 1800, on the effect of mass on tendencies of chemicals to combine.

SUMMARY

The analyses presented in this chapter provide evidence for both the set of hypotheses developed in Chapter 2 and the theory behind them. Actually, aside from details of dates, only two hypotheses were disconfirmed: (H17) that use of affinities was associated with quantitative papers, and (H19) that French

[17]Due to Berthollet's work in particular, the use of affinities was positively correlated with oxygen theory from 1785 to 1795 ($r = .19$) in France.

articles were more theoretical than British. The latter hypothesis was more of a conjecture, not based on the theory, and we accounted for its failure in Chapter 4, while the evidence against the former probably indicates that affinities proved difficult to quantify and were used by many chemists to account for purely qualitative results.

The following hypotheses were supported: (H1) the proportion of articles using the oxygen paradigm increased over time, while the proportion using phlogiston decreased; (H2) the proportion of theoretical articles increased until the climax of the paradigm dispute, 1785-88, then decreased, with articles using oxygen being initially highly theoretical; (H3) the proportion of quantitative articles increased over time, but the largest increases were before 1778 and after 1788; furthermore, the proportion of quantitative papers was initially very high among those using the oxygen paradigm and declined steeply among phlogiston articles after 1785; (H4) the more theoretical the article, the more likely it was to be quantitative (but this relationship was true even before 1777); (H5) that French articles would be much more likely to use oxygen than British; (H6) that theoretical papers would be more likely to use oxygen than nontheoretical ones, although the measured relationship was weaker than it might have been, due to the phlogistonists defending themselves in theoretical terms; (H7) that quantitative papers would be more likely to use oxygen than qualitative ones; (H10) that the number of neutral papers would be very small; (H11) that French articles would shift to use of the oxygen paradigm almost a decade before the British; (H12) that the shift to the oxygen paradigm would occur earlier among theoretical papers than among nontheoretical ones; (H13) that the same shift would occur earlier among quantitative papers than among qualitative ones[18]; (H14) that younger authors would switch to the new paradigm earlier than older ones with the exception of Lavoisier and Bayen; (H15) that the younger an author the more likely he would be to use the oxygen paradigm; (H16) that the use of affinities would be positively associated with theoretical papers; and (H20) that French papers would initially be more quantitative than British (although the difference did not increase as predicted as the Revolution progressed).

The data further show that, given anomalies, levels of theory and quantification increased, and theoretical or quantitative work led in turn to questioning of the given paradigm and the proposal of a new one. As the crisis grew, the effect of quantification became especially acute. Finally, as the new paradigm won adherents and came to dominate the field, passing time and age were the variables which made the most difference. We further saw that theory and quantification interacted in such a way that it was exceptionally difficult for papers which were both theoretical and quantitative to avoid the revolutionary

---

[18]These last two results were reversed for the British papers, but since the shift had barely begun and few papers were involved, we do not feel that this is a significant counter-example.

implications of anomalies; this, however, was less the case for papers either non-theoretical or qualitative, or both.

Finally, some new and unexpected findings emerged. We discovered that British chemists as a group became older until late in the Revolution, while French authors became younger; that is, young men were attracted to the field in France but not in Britain. This fact, coupled with the expected finding that young scientists were more likely to support the new paradigm, provides a new insight into the well-known resistance of British chemists to the oxygen paradigm: much of their resistance was due to their age and its natural conservatism.

We now turn to a consideration of recognition and a further investigation of the structure of the chemical communities of eighteenth century France and Great Britain. This analysis gives us further insight into and support for the argument that nonrational factors played an important part in the progress of the Revolution, an argument that has already been enhanced by the findings that nationality, age, nomenclature, and personal contact with supporters of the new theory all had a strong impact on the acceptance and dissemination of the new world view.

6

Recognition and the Community

Now that we have examined the theory in detail and found it useful for predicting the course of scientific revolutions, we will return to the discussion of community structure begun in Chapter 3. Since one of our basic assumptions is that
scientists are motivated by a desire for recognition, and particularly formal
recognition, through the same channels that provide for communication, we will
now look more closely at the citations. As noted in Chapter 1 (p.11), we are
utilizing citation data in three principal ways: (1) as an indicator of revolutionary
shifts, specifically to examine the shift to the oxygen paradigm (H18); (2) as
an indicator of communities; and (3) as an approximate measure of prestige.
We shall study overall citations, citations to the authors in our sample, and
comparisons between countries. This should give us a more detailed picture of
the social organization of the chemical community at this critical juncture in its
history, a time not only of the birth of a new paradigm, but of the emergence of
chemistry as a profession.

Table 6.1 shows that French authors cited countrymen about 50% of the time,
somewhat more in the earlier periods (especially 1760-71), less during the peak
debate periods (1781-88), and more again during consolidation of the new
paradigm (1789-95). Citations to British authors averaged only 10%, more during
the period when pneumatic chemistry spread from Britain (1772-80) and in
1787-90. The remaining 40% were to authors of other nationalities.[1]

British authors, on the other hand, cited French authors 27% of the time,
British authors 36%, and other nationalities 37%. Thus, relative to French
authors, the British cited their own countrymen more and Frenchmen less, with
about the same proportion of citations to others. They relied on the French
more than the French relied on them, both overall and in all time periods after

---

[1]Some of this 40% is guesswork; we do not know the nationality of all of the scientists
cited.

TABLE 6.1

Percentage Distribution of Citations by Nationality of Those Cited over Time

To	1760 -65	1766 -71	1772 -77	1778 -80	1781 -84	1785 -86	1787 -88	1789 -90	1791 -95	Total
					By French authors					
French	57	57	53	52	48	48	42	52	53	50
British	1	5	14	12	7	11	14	12	9	10
Other	42	38	33	36	44	40	45	37	38	39
Total	100	100	100	100	99[a]	99[a]	101[a]	101[a]	100	99[a]
Number of citations	(103)	(92)	(347)	(225)	(397)	(332)	(237)	(205)	(319)	(2257)
					By British authors					
French	0	15	9	36	27	30	31	29	31	27
British	100	56	55	27	32	21	40	39	30	36
Other	0	30	36	36	41	49	29	32	39	37
Total	100	101[a]	100	99[a]	100	100	100	100	100	100
Number of citations	(1)	(27)	(44)	(11)	(95)	(57)	(42)	(87)	(83)	(447)

[a] = rounding error

1777; only in 1772-77 were the French more dependent, as British pneumo-chemistry swept Europe. It is hard to discern any clear patterns in the distribution over time, except that citations to French authors were slightly greater over the last ten years than earlier.

Since we have almost no information (including nationality) about authors outside our sample, either those in other countries or from earlier periods, in order to be precise about numbers, we will limit our attention to citations to authors within the sample. Table 6.2 shows that of the 1032 citations by French authors, 84% (866) were to Frenchmen and 16% (166) were to British. Further, these 866 citations constituted 76% of all their citations to French authors and 72% of all their citations to British authors. It is interesting to note that the proportion of citations to British authors was greatest (22-25%) in 1787-90, when the new paradigm was making its greatest gains among French chemists, probably as a result of attacks on British phlogistonists. Overall, French chemistry was quite French.

British citations to authors in the sample were almost equally divided, with somewhat more to British chemists (54%). Again the distribution over time is erratic, and patterns are hard to discern in French-British differences, but there does seem to have been a slightly higher proportion of citations to French authors during 1778-86 (averaging over 50%) — probably a product of the developing French interest in pneumatic chemistry and the challenge to the British phlogistic explanations by Lavoisier. Again in 1791-95 there was a rise in citations to French authors (53%), probably a result of the conversion of British chemists to the new paradigm.

The proportion of citations to *all* French and British authors which were to authors *in the cited sample* also provided some interesting information. We note two facts: first that the proportion of citations by British authors to chemists in the sample (87% of those to French authors and 75% of those to British authors) was higher than that of French authors (76% of those to French authors and 72% of those to British); second, that the proportion of citations *to* French authors in the sample was higher (76% from French chemists and 87% from British) than the proportion to British (72% from French authors and 75% from British). We can think of this as a measure of the timeliness of these citations, since other citations to countrymen were, with few exceptions, to men who published earlier (than 1760).

In this sense citations *by* British authors were more current (higher proportion to their contemporaries) than those by French authors, while citations *to* British authors were less current than those to French authors. This may well reflect both the great growth of chemistry at this time and the fact that British authors had less of a tradition in chemistry. This reasoning is supported, we think, by the smooth growth of this proportion in the French data, by the extremely high (in relative terms) proportion of British authors in the last eight years, when British chemistry began to develop more, and by the relatively sharp break in

TABLE 6.2
Citations to Authors in the Sample[a]

	1760 -65	1766 -71	1772 -77	1778 -80	1781 -84	1785 -86	1787 -88	1789 -90	1791 -95	Total
					By French authors					
% to French	100	100	87	81	89	82	75	78	84	84
% of FR	41	54	62	74	87	89	87	75	82	76
% to British	0	0	13	19	11	18	25	22	16	16
% of BR	0	0	35	74	72	82	88	96	93	72
Total %	100	100	100	100	100	100	100	100	100	100
Number	(24)	(28)	(131)	(106)	(188)	(174)	(114)	(103)	(164)	(1032)
					By British authors					
% to French	0	28	13	100	48	62	38	44	53	46
% of FR	0	50	50	75	88	94	77	92	96	87
% to British	100	72	87	0	52	38	62	56	47	54
% of BR	100	33	54	0	83	83	94	85	88	75
Total %	100	100	100	100	100	100	100	100	100	100
Number of citations	(1)	(7)	(15)	(3)	(48)	(26)	(26)	(52)	(47)	(225)

[a]The lines marked "% to" are the relative percentage of citations to French or British authors within each of the citing groups (French or British). The lines marked "% of FR (BR)" are the proportion of citations to FR (BR) are the proportion of citations to authors in the sample relative to citations to all authors of the same nationality (that is, relative to the respective numbers in Table 6.1).

TABLE 6.3
Citations to Oxygen Authors over Time

	1760 -65	1766 -71	1772 -77	1778 -80	1781 -84	1785 -86	1787 -88	1789 -90	1791 -95	Total
				By French authors						
Oxygen	0	0	13	5	16	47	40	50	105	276
% of CA^a	0	0	10	5	9	27	35	49	64	27
				By British authors						
Oxygen	0	0	0	0	9	5	6	13	19	52
% of CA	0	0	0	0	19	19	23	25	40	23

^{a}CA = total citations to authors in the sample (by the appropriate group).

the proportion of French citations to French in 1781-84, as the old paradigm came under sharp attack. There was a particularly high concentration of current citations in the last ten years of the sample, a figure clearly to be expected in the light of a paradigm shift which renders most of earlier work obsolete (see Chapter 1, p. 20).

Of particular note is the fact that citations by British chemists to French authors, either those in the sample or not, were very scarce during the first 18 years (1760-1777). It was presumably from French chemists, particularly Macquer, that the British picked up phlogiston theory (Schofield, 1970, p. 191), but the data do not support this view.

We now turn to the distribution of citations to oxygen chemists and a test of (H18), that, at the expense of the phlogistonists, the proportion of citations to chemists subscribing to the oxygen paradigm will increase over time.[2]

The data generally support (H18), although the date when oxygen citations became "dominant" depends on the definition of the word and on what method of calculation one uses to compute the percentage. Fifty per cent seems to us a

[2] Oxygen citations were calculated in this way: an author was considered to be in the oxygen group once he had published an article using the oxygen paradigm. This results in two sorts of imprecision, which tend to cancel one another: (1) it treats citations to non-oxygen articles by oxygen authors (those articles written before the conversion) as "oxygen" citations, thereby becoming more a measure of the relative dominance of authors who switched than a measure of the relative dominance of the oxygen paradigm. This is hard to avoid, since very few citations were to individual articles by name — one might be able to figure out the exact citation in many or even all cases (though we doubt it) — but this would require years of work hardly worth the effort in a preliminary study. (2) Since we wait until an author has published an article using the oxygen paradigm before putting him in the oxygen camp, there may be some citations to work using the oxygen paradigm not counted as "oxygen" citations. Again, this would be difficult to determine. As explained in Appendix A, it seems to us to be better to be internally consistent — to use publishing dates throughout — than to try to determine whether references are to unpublished work or to earlier work.

reasonable (albeit arbitrary) level for "dominance," and our choice of calculating methods is one which uses the proportion of all citations to authors in the sample (since we do not know the paradigm choices of authors outside the sample). Thus, the 50% level was reached in 1791-95.[3] Therefore, oxygen chemists were dominant in terms of prestige (citations) by about 1790 in France.[4]

Finally we will consider the relative importance of members of the Académie des Sciences and the Royal Society for French and British authors respectively.[5] The data are in Table 6.4.

It is clear from the table that Academicians were important for French chemistry, receiving 62% of all citations by French authors to French authors in the sample,[6] a percentage which fluctuated only a few points over all time intervals. It was slightly lower in 1772-80, when new chemists were being recruited rapidly (see Chapter 3) and the *Observations sur la Physique* provided a ready vehicle for their production. The percentage was somewhat higher (71%) in 1760-65, when major journals were controlled by the Académie, and again (70%) in 1787-88, just after the conversion of several Academicians, such as Berthollet and Fourcroy, to the new paradigm.

The proportion of citations to Academicians was greater than the proportion of Academicians or articles by Academicians in every time period, often by a large margin.[7] For example, during the important years of 1785 to 1790, the proportion of Academicians ranged from 30 to 56%, their fraction of articles

[3]Looking at data for individual years, 50% was reached in 1790 and a peak of 88% in 1794.

[4]The proportion of citations to authors subscribing to the oxygen paradigm lagged quite a bit behind the proportion of authors and the proportion of articles which used the oxygen paradigm, as a comparison with Chapter 4 will show. This indicates that the first type of imprecision spoken of in footnote 2 is probably not very serious.

[5]There were no British members of the Académie, but there were French members of the Royal Society, which fact we have ignored as Académie membership was far more important.

[6]When we count *all* citations to these men, instead of just those after their election, they appear even more dominant. Working Academicians elected in 1795 or earlier received 76% (656) of the citations by French authors to French authors in our sample and 77% (80) of all the citations by British authors to French authors in the sample. On the other hand, if we restrict ourselves to "chemists" (members elected into or moved into the section of chemistry), we find that the 16 men elected before 1795 received (after they were elected) 53% (455) of all citations by French authors to French authors and 58% (60) of the citations by British authors to French. The 19 chemists elected in 1795 or earlier received a total of 63% (548) of all French citations to French authors in the sample and 70% (73) of all British citations to French authors in the sample. Thus a mere handful of men working in or near Paris not only contributed more than their share of articles (see Chapter 3), but received most of the recognition (in the form of citations).

[7]See Table 3.9 for the comparisons.

TABLE 6.4
Citations to Members of the
Académie des Sciences and the Royal Society

	1760 -65	1766 -71	1772 -77	1778 -80	1781 -84	1785 -86	1787 -88	1789 -90	1791 -95	Total
					French authors					
Académie	17	19	68	49	104	95	60	48	84	540
% of FR[a]	71	68	60	57	62	66	70	60	61	62
Royal Soc.	0	0	8	16	18	29	26	21	24	141
% of BR	—	—	47	80	86	94	89	91	92	85
					British authors					
Académie	0	0	2	2	16	7	8	15	17	67
% of FR	—	0	100	67	70	44	80	65	68	64
Royal Soc.	0	3	10	0	19	7	16	25	17	97
% of BR	0	60	77	—	76	70	100	86	77	80

[a]FR = citations to French authors; BR = citations to British authors. For Academicians the table is restricted to citations *after* they were elected and is also restricted to "working" members, that is, those who were regular members of one of the sections and not honarary or "libré." This makes the table comparable to Table 3.9 in Chapter 3.

ranged from 47 to 69%, while their share of citations ranged from 57 to 70%. The difference between the proportion of articles contributed and the proportion of citations received ranged from 1% (1787-88, 1789-90) to 17% (1785-86). This we would expect, given the assumed superiority of Academicians to other French chemists. It may also represent a halo effect; that is, Academicians may have been cited because they *were* Academicians and not because they produced more important work. However, it is unlikely that this factor could account for the magnitude of the difference between their number and their relative numbers of citations, particularly since they did contribute disproportionately in quantitative terms. Quantity is not equivalent to quality, but it does indicate a strong commitment to the field, especially with no "publish or perish" syndrome working.

Not only does this indicate the use of citations in determining prestige, but we also begin to get a sense of community: a relatively small group of scientists were relied upon by a larger group. It was not a case of random citations.[8]

While their countrymen cited Fellows slightly more overall than their proportion of the population[9] (80 to 78%), they did so in only two time periods

[8]The prestige aspect of citations has a hint of circularity here: we could either be saying that because of the large fraction of citations these men received, they acquired great prestige, which was formally recognized by election to the Académie, or that we know the citations reflect prestige because Academicians do, in fact, have high prestige.

[9]Again, the numbers for the comparsions may be found in Table 3.9.

(100% in 1787-88 and 86% in 1789-90). Furthermore, Fellows received less than their share of citations relative to their articles in five of the nine time periods: 1760-65 (50 to 0), 1766-71 (80 to 60), 1778-80 (80 to 0), 1785-86 (100 to 70), and 1789-90 (95 to 86). It is not clear, then, whether or not their work was better than that of their competitors. In any case, the number of citations they received does not stand out in the way that the Academicians' did.[10]

The tables here and in the first section have shown the distribution of papers and citations over time by various general categories. We now wish to consider a more detailed picture of the top men — the major authors and citation receivers. The productive authors contributed more than their share of information to the community and should, therefore, have received more than their share of the recognition.[11] Of course, this assumes some relation between quantity and quality, an assumption not always justified. For example, Black published only three articles in his lifetime, but he is universally regarded as an important and influential chemist. Too, in contrast, publication of many poor articles may result in very few citations.[12] Furthermore, while books were more important as original contributions among the sciences of the eighteenth century than among the more developed sciences of today (see Chapter 3, p. 52), we have not considered any monograph literature, a strategy which may have led to minor omissions.

The lists of prolific producers and of citees are similar (the overall association between number of articles and number of citations is .64 (Pearson's r)) and contain many of the same names, especially for Britain. However, they are even more similar than they first appear for some of the discrepancies are only superficial and others are easily accounted for. For men on the top producers list not on the top citees list, Pelletier ranked 15, Cornette 21, Lassone 11, and Hassenfratz 20, leaving only two: Vauquelin, who was very young, and Lamétherie, who was rather peripheral to the field, being a journal editor and natural historian (as well as an opponent of the new theory). Of those who received many citations but did not publish very many articles, Macquer was a chemist of an earlier era and ranked 14, Baumé ranked 16, Monnet 12, and Cadet 11. To go further down

[10]However, one could argue that it would be difficult for them to stand out because of a "ceiling effect." The fact that they were such a large fraction of British authors to begin with makes it very difficult for them to have received an even larger fraction of citations. However, since their proportional representation was higher among citations by French authors, they could have been even better represented among citations by British authors.

[11]This is an obvious assumption of the "information-recognition" model developed by Hagstrom (1965, pp. 12-13) and noted in Chapter 1, p. 9.

[12]One would expect this to be less of a problem in the eighteenth century than it is now, since there was no "publish or perish" syndrome at work. Curiously, however, although Black was very influential (Guerlac, 1961a), he published very little and was never elected a Fellow of the Royal Society. He was, however, a notable teacher.

the list, 19 of the top 20 French producers of articles were among the top 20 French receivers of citations and 19 of the top 20 French receivers of citations were among the top 20 producers of articles. Thus, there is quite a bit of correspondence between the two sets of rankings as predicted.

Noting the positive relation between quantity and quality, we would expect Academicians (and, to a lesser extent, Fellows) to be both prominent (overrepresented) in articles produced and frequently cited — if, in fact, the Académie selected well. We refer here to the most respected chemists; we already know that Academicians, at least, were disproportionately prominent in producing articles and receiving citations.

What do the data show? Restricting ourselves to citations to authors in the sample, we find in Table 6.5 that members of the Académie and the Royal Society dominate these lists: Nine of the eleven most productive French authors were Academicians, as were nine of the ten most frequently cited authors, while for Britain all of the top five producers (true for the French as well) and four out of five of the most frequently cited authors were Fellows. This is further and convincing evidence of the relative dominance of the Académie in spite of its restricted membership.[13]

The relative dominance (that is, the proportion of articles produced or citations received) of the outstanding authors is also of interest. It is surprising how a few individuals can dominate a field, as shown in Table 6.6.

We see that a very large number of the citations were going to very few chemists. While the top five chemists constituted only 3.9% of all authors, they received 34% of all citations, 35% of the citations by French authors, and 45% of the citations by British authors.[14] Likewise, the top ten constituted 7.8% of all authors but received 53% of all citations, 50% of those by French authors, and 61% of those by British authors. In the important decade of 1781-90, the top five citees received 36% of all citations and the top ten received 54%, a slightly greater concentration than for the total years.[15]

---

[13] Also interesting are the men on these lists who were not Academicians. Both Lamétherie and Monnet were opponents of Lavoisier (Rappaport, 1969) and never converted. Thus, Hassenfratz is, perhaps, the only surprising omission, and Partington (1962, p. 410) has little regard for his work. As regards Fellows of the Royal Society, the case of Black has already been mentioned (footnote 12). To go further down the list, 15 of the top 20 French authors (in terms of articles produced) were Academicians and another became one by the end of the century. Also, 15 of the top 20 French citees were Academicians with another becoming a member by the end of the century (Chaptal in 1796).

[14] Note that these are not necessarily the *same* five authors (or ten) in each case: the top five receivers of citations by French chemists were Macquer, Lavoisier, Priestley, Baume, and Berthollet, in that order; of citations by British chemists, Cavendish, Priestley, Lavoisier, Kirwan, and Black; of all citations see Table 6.5.

[15] The top ten in order: Priestley (65), Lavoisier (62), Macquer (51), Berthollet (45), Cavendish (35), Baumé (33), Morveau (25), Kirwan (24), Monnet (21), and Sage (20).

TABLE 6.5
Rankings of the Top Ten French and
Top Five British Authors by Publications
and Citations

Publications[a]			Citations[a]				
					Number		
Authors	Number	%	Authors	FR[b]	BR[b]	T[b]	%
French							
Sage[c]	61.3	10.0	Lavoisier[c]	76	22	98	7.8
Lavoisier[c]	53.7	8.7	Macquer[c]	88	6	94	7.5
Guyton[c]	42.8	6.9	Baumé[c]	71	8	79	6.3
Berthollet[c]	39.8	6.5	Berthollet[c]	55	8	63	5.0
Fourcroy[c]	36.9	6.0	Guyton[c]	35	8	43	3.4
Pelletier[c]	27	4.4	Sage[c]	33	2	35	2.8
Cornette[c]	21.3	3.5	Fourcroy[c]	31	3	34	2.7
Lassone[c]	21.3	3.5	Darcet[c]	32	1	33	2.6
Vauquelin[c]	21.1	3.4	Monnet	25	5	30	2.3
Hassenfratz	20	3.2	Monge[c]	24	1	25	2.0
Lamétherie	20	3.2					
British							
Priestley[c]	21	20.8	Priestley[c]	73	24	97	7.7
Kirwan[c]	14	13.9	Cavendish[c]	19	26	45	3.6
Cavendish[c]	9	8.9	Black	18	14	32	2.5
Woulfe[c]	9	8.9	Kirwan[c]	16	15	31	2.5
T. Henry[c]	7	6.9	Woulfe[c]	10	1	11	0.9

[a]For publications, % = percentage of total number of articles by countrymen (that is, 616 French articles or 101 British articles). For citations, % = percentage of *total* (British plus French) citations to authors in the sample (1257).

[b]FR = citations by French authors; BR = citations by British authors; T = total citations.

[c]Members of Académie or Royal Society. The decimal numbers for the French authors are due to multiple-authored articles, which were divided equally among their authors.

Once again, this shows us the usefulness of citations in identifying communities (in a crude sense of the word): these results give us a picture of a small community of chemists responsible for most of the work and who attain most of the recognition in the field, surrounded by a much larger group of relatively minor figures.

While we are looking at the top citees in the literature, we should not ignore chemists outside our sample since several were influential and received a large share of the citations. The most-cited chemist (126 citations) was Torbern Bergman of Sweden, who wrote the definitive work on affinities (Guerlac, 1961a, p. 208). Other important men were Scheele (95) ("co-discoverer" of the gas now called oxygen), the German chemists, Marggraf (64) and Pott (50), and the famous French teacher of chemistry, G. F. Rouelle (56), who published

TABLE 6.6
Proportion of Citations to Top Chemists[a]

Citations to	Citations by			1781-90 (all)
	All	French	British	
Top five	34.3	35.2	44.9	36.2
Top ten	53.4	50.3	60.9	53.8
Top twenty	65.1	68.3	–	70.1
Total number	1257	1032	227	730

[a]The table gives the percentages of citations to the top receivers by all authors, by French authors, and by British authors. Of the 730 citations in 1781-90 (58% of all citations to authors in the sample), 577 were by Frenchmen (56% of all citations by French chemists) and 151 were by British (67% of all citations by British chemists).

only five articles in his life. Other authors received considerably less attention, except for the founder of phlogiston chemistry, Stahl, who was cited 47 times, only three by British chemists.[16] The French, having had a strong section of chemistry in the Académie since the beginning of the eighteenth century also cited with some regularity two Academicians of bygone days, Geoffroy (42) and Lemery (31). In all, there were 25 men outside the sample who were cited ten or more times by French chemists, and of these, ten (40%) were German, six were French, two were British, and two Scandinavian.

The strong representation of Swedish and German names, the latter particularly among French chemists, was probably due to the French interest in mining and metallurgy (an interest not found among the British) and to the development of phlogiston theory in France (Guerlac, 1959a; Rappaport, 1961). Bergman was especially important because of his work on affinities, and of special interest to both Berthollet (who cited him 17 times) and Kirwan (10 citations).

Furthermore, there was a close relationship between the French chemists and those at Berlin (such as Marggraf and Achard), where the *Mémoires* of the Academy of Sciences were published in French throughout most of the eighteenth century.[17] Schofield (1970, p. 191) also noted the ties between British and Continental chemistry, particularly the Germans.

G. F. Rouelle, to whom we have frequently referred, provides an interesting and informative sidelight. We have noted, following Rappaport (1960, 1961),

[16]This may seem less unlikely if we say that we did not count references to "Stahlians" or "Stahlian chemistry" but only to Stahl's work, just as with the other authors.

[17]The close relationship was noted by the pharmacist and demonstrator in chemistry, Mitouard (friend of Lavoisier), in a letter to the author of the *Observations sur la Physique. . .* 2, 323-324 (October, 1773). Checking the Berlin *Mémoires* for the years 1760-95, we found the following: there were 116 citations, 23 (20%) were to Frenchmen, 10 (9%) were to Englishmen, and the remainder were almost exclusively to Germans and Scandinavians. Among the top citees were Pott (ten) (a favorite of Marggraf and Achard, who published 13 and 49 articles, respectively, during this period, dominating the Berlin Academy), Marggraf (seven), Priestley, Bergman, and Scheele (five each). Lavoisier received only two.

that he was a very influential teacher and the man most responsible for populariz-ing phlogiston theory in France and, through Macquer's writing, Great Britain (see also Mayer, 1970). He is particularly important because he was Lavoisier's teacher. But his influence did not stop there. Directly or indirectly, he taught or influenced almost all of the significant chemists in France, those who supported oxygen and those who resisted. His students included Lavoisier, Macquer, Sage, Cadet, Proust, Darcet, H. M. Rouelle, Baron, Bayen, Brongniart, Bucquet, Montet, Roux, Venel, Monnet, and the Englishman, Woulfe. In addition, Baumé was an assistant to his student, Macquer; Berthollet and Fourcroy studied with Macquer and Bucquet; Dize and Pelletier studied under Darcet; and Guyton de Morveau learned his chemistry from Macquer's text (which was also used by Cullen and Black at Edinburgh).

Referring to Tables 6.5 and 6.6, we see that among the top 11 French authors (number of articles), the top two were Rouelle's students and another six learned from his students (Hassenfratz worked with Lavoisier); among the top ten French citees, the top two were his students, as were three others, and an additional four were students of students (including Morveau). His student Woulfe was among the top five British authors and citees. Among the top five receivers of French citations, only Priestley was not one of his students. This would seem to be prima facie evidence that the best way to become a well-known chemist in the late 1700s was to study with Rouelle — even though he published little, his students were very prolific. They were also honored, for in addition to their representation among the major receivers of citations, most became members of the Académie, the only notable exception being Monnet (an opponent of Lavoisier). Darcet became the first president of the Académie not appointed by the king.

Finally, and ironically, Rouelle, the primary promoter of phlogiston, trained the earliest opponents of that theory: Lavoisier, Bayen, Bucquet (Lavoisier's earliest follower), and Turgot.[18] To compound the irony, students of his most prominent disciple, Macquer, included such early converts as Berthollet and Fourcroy, as well as the man who converted when he was over 60, Baumé. So this one man was influential, either directly or indirectly, in two of the major revolutions in chemistry.

SUMMARY

The data showed strong support for the hypothesis (H18) that citations to oxygen chemists would increase at the expense of citations to phlogiston chemists. The proportion of citations to oxygen chemists, however, did lag behind the

[18]Turgot, although not a scientist, was a strong supporter of the sciences and wrote some reflections in which he expressed doubt about the phlogiston theory. We will return to this in the next chapter.

proportion of articles supporting the oxygen paradigm, showing that the defenders of phlogiston were able to put up a fight. We also found that French chemists were generally the men who were recognized by the community through citations, although the most prolific British chemists were also among the leaders. It is interesting to note that there were men such as Black and G. F. Rouelle, who wrote very little but who received great recognition. While this could be an indication that quality is more important than quantity (the same could be said for the relative positions of Sage and Lavoisier on the two scales of production and recognition), it seems likely that it is also a demonstration of the effects of good teaching, for both were celebrated teachers.

As in the case of productivity, the Paris Académie was overrepresented among the top receivers of recognition. Although this may be partially due to a halo effect, it is more likely caused by the greater quality of the work done by Academicians, reflecting both their careful selection for membership and their better command of the resources of the scientific community, including laboratories, salaries, and high status. An Academician received all the rewards that a scientist could want, including respect from colleagues and the public.

We also began to look at some of the highly cited men outside the sample, finding that Germans and Scandinavians received the bulk of recognition. Bergman and his student, Scheele, were by far the most frequently cited "outsiders," and Bergman was the most frequently cited chemist of all, including Lavoisier and other prominent Academicians. Since we are assuming citations to be an indication of prestige, we would have to say that Bergman was the most respected chemist in Europe, if not the most important.

Finally, we saw three uses of citations: (1) to show a shift to the supporters of the new paradigm; (2) to show that there existed a community of chemists, consisting of a small group of frequently cited men surrounded by a much larger group of peripheral chemists – there was also evidence of two national sub-communities, French and British, shown by differences in the frequency of citation within and between these two and by the differing frequencies of citations to supporters of the new paradigm; and (3) to show that there was a small group of chemists with high prestige as regard citations, reflected in France by their high rate of election to the Paris Académie des Sciences. We also saw that G. F. Rouelle, whom Rappaport feels is a neglected figure (1960), was in several ways the most influential chemist of the period.

This citation analysis may also shed light on the issue of risk-taking and conversion to a new theory. In Chapter 1 we argued that new members, either young or from other fields, are more likely to take the risk of supporting a new theory because they have little to lose and much to gain. That view seems to be supported by the data in France: younger men converted earlier, as seen in Chapter 4 and 5, and in this chapter we learned that those who converted early to oxygen theory received a disproportionate share of the recognition. However, it is also apparent that prominent but not especially young men, such as Berthollet,

Fourcroy, and Morveau, were also able to perceive and accept the new view rapidly. This appears to stem from their personal contact with Lavoisier along the lines of "contagion" models (cf. Crane, 1972, Chapter 2).

We also found in Great Britain, however, that younger chemists did not accept the new paradigm very rapidly. The data in this chapter suggest that this may have had something to do with their deference to prominent British chemists: unlike France, the most prestigious chemists in Britain, as measured by citations (Priestley, Cavendish, and Kirwan — Black is somewhat of a special case, since he was neither publishing nor clearly on either side according to the historical accounts) and by the secondary historical literature, were strong supporters of the phlogiston theory. Therefore, it may have been too risky for new young chemists to publish in support of the opposing French chemistry. A contributing factor was probably that these chemists could see the experiments of a Cavendish supporting phlogiston theory but were not able to see or participate in the experiments of the oxygen group.

The citation data, then, provide another example of nonlogical factors in the acceptance of the new paradigm: men who achieved prominence heavily influenced the choices of their colleagues. Those who worked or trained together tended to have similar views. Finally, the importance of certain influential teachers appeared in the citations and careers of Rouelle and his students.

7

Revolution Revisited

The preceding four chapters have presented various tests of a system of twenty hypotheses, uncovered some new and unexpected historical facts, and suggested some explanations for these facts and for the failures of the theory. In this chapter I will first summarize the findings relative to the test of the theory and models, then discuss the relevance of the results for the theories of Kuhn and Hagstrom and the implications for the sociology and history of science. Finally, I will offer some speculation about the status of Lavoisier.

STATUS OF THE THEORY

The central and most important part of this study is the strength and usefulness of the theory developed. In general, we can conclude that the theory is strongly supported by the data in various forms, and that it is useful for investigating the processes of scientific change and revolution. With very few exceptions the hypotheses generated by the general theory have been confirmed, although one case study cannot be conclusive. Furthermore, the causal models based on the theory seem to reflect the underlying processes of change and revolution. The basic model and the overall model for each country explained from 40 to 50% of the variance in paradigm choice, a gratifyingly high amount, especially for such a small set of predictor variables.

On a general level, our data have clearly shown that nonlogical or nonscientific factors played an important role in the development and acceptance of the new paradigm. Specifically, we found that nationality and age were significant causes of paradigm choice, more important than any other variables except passage of time. To document this we will take each hypothesis in turn and assess its standing.

(H1) The proportion of oxygen articles will increase over time and the proportion of phlogiston articles decrease. This hypothesis, a particularly obvious one, was supported by the data of Chapters 4 and 5, though there was little evidence for it in the British data of this period.

(H2) The proportion of theoretical articles will increase until some time between 1785 and 1789, then decrease; the proportion of theoretical papers will be high among oxygen papers at first and decline as the oxygen paradigm becomes accepted. Both parts of this hypothesis were clearly supported for the French chemists by the data in Chapter 4 and 5, while the latter part was essentially irrelevant for the British.

(H3) The proportion of quantitative articles will increase, especially after 1777; the proportion of quantitative articles among oxygen articles will be very high at first, and the proportion of quantitative articles among phlogiston articles will decline after about 1785. The first part of this hypothesis was supported for both countries by the data in Chapters 4 and 5, but not very strongly: there was a leveling off and perhaps even a decline, attributable to the presence of new problems not amenable to the kind of quantification we are measuring. The second part of the hypothesis was strongly supported by the French data, but not by the British.

(H4) The more theoretical the articles are, the greater the proportion will be quantitative after 1777. This hypothesis was supported only in part by the data of Chapter 5, partly because of the artifacts of the data, and partly because the true relationship may well be curvilinear.

(H5) There will be a larger proportion of oxygen articles among the French articles than among the British at any point in the Revolution. This was overwhelmingly supported by the data of Chapters 4 and 5: there were almost no British articles supporting the new paradigm.

(H6) The proportion of papers using the oxygen paradigm will be higher among the more theoretical papers than among the less theoretical. This hypothesis was strongly supported by the French data of Chapters 4 and 5, but not the British.

(H7) The proportion of papers using the oxygen paradigm will be higher among quantitative papers than among qualitative ones. Again the data of Chapters 4 and 5 strongly supported this hypothesis for French articles but were inconclusive for British.

(H8) The number of authors will increase rapidly after 1777 (and may also increase rapidly after 1772 due to the great interest aroused by the new gas chemistry). (H9) The productivity of the authors will increase as the Revolution progresses, especially after 1777. Both of these hypotheses were supported by evidence presented in Chapter 3. The details of the variations in productivity over time fit our hypotheses, although the *overall* trend of growth may be due to factors peripheral to the Chemical Revolution.

(H10) The number of neutral papers will be small. This hypothesis was supported by the data of Chapter 5, although its meaning is somewhat problematic.

(H11) French articles will shift to the oxygen paradigm before British ones.

(H12) The shift to the oxygen paradigm will occur earlier among the more theoretical papers. (H13) The shift to the oxygen paradigm will occur earlier among quantitative papers than among qualitative ones. (H14) Younger authors will shift to the oxygen paradigm earlier than older authors. All of our shift hypotheses were overwhelmingly supported by the French data. The British data appeared mixed but were so sparse that they led to no conclusions. Even the anomalies, the French diehards, were predictable from the theory: they were older, their papers less theoretical and qualitative.

(H15) The younger an author is, the more likely it will be that he will use the oxygen paradigm once the choice is available. The data show that this is clearly the case, and we suggest that this finding, together with the age difference between French and British chemists, explains much of the resistance to the new paradigm in Britain.

(H16) There will be a positive association between theoretical papers and use of affinities. This was supported, although the cause may have been partly the process of classification. The changing correlations argue against this view.

(H17) There will be a positive association between quantitative papers and use of affinities. This was one of the few hypotheses not supported, probably because of the failure of affinity theorists to quantify chemical attraction, as well as the irrelevance of weight for affinity theory until Berthollet's work after 1800.

(H18) The proportion of citations to authors who supported the oxygen paradigm will increase, and the proportion of citations to authors who used phlogiston will decrease as the Revolution progresses, especially after 1785. This hypothesis was supported by the data of Chapter 6, and it, too, is obvious.

(H19) The proportion of papers which were theoretical will be higher among French than among British articles, and this difference will increase as the Revolution progresses. This hypothesis was the only one strongly falsified: the British articles were significantly more theoretical than the French. We suggest, however, that this was due, paradoxically, to the greater professionalism of the French scientific community, allowing and encouraging publication by workers who, in Britain, would have been too preoccupied trying to make profits.

(H20) The proportion of papers which were quantitative will be higher among French than among British articles, and this difference will increase as the Revolution progresses. This hypothesis was not supported in the sense that there was very little difference between the two countries.

Some additional data which was not presented during the examination of the path models indicates that certain of our assumptions may be suspect. As noted, we assumed that any article classified "old nomenclature" was a "phlogiston" article and that any article classified "new nomenclature" was an "oxygen" article (although we point out in Appendix A that there was at least one discrepancy from the first assumption in the British data). Furthermore, we also noted that the five "phlogiston" articles by men under 40 after 1736 were classified due to "old nomenclature," although we did not stress the fact that this allows for the possibility that they were not, in fact, "phlogiston." We do feel that our theory, particularly the part that makes use of the gestalt aspect of paradigms, must assume that that these classifications are proper: if there is, in fact, a gestalt, then one cannot accept the paradigm and reject the nomenclature, or vice versa. Thus, in the French data there is the possibility, and in the British data the certainty, that the oxygen theory did not function as a gestalt for at least a few chemists. Nevertheless, we found no evidence to counter the assumption that acceptance of the paradigm meant acceptance of the nomenclature.

There is other relevant evidence. Not only did the phenomenon of fragmentation of the old paradigm during a crisis state occur, but there is also evidence that some scientists accepted parts of both paradigms in this particular case. Crosland (1973, pp. 322-323) points out that there were several chemists in the 1780s who maintained allegiance to the phlogiston theory while still accepting Lavoisier's explanation of acidity (that acidity was caused by the presence of oxygen), an explanation which opposed the phlogistic view of acidity or causticity.

Perhaps we can construe these phenomena as temporary aberrations caused by the psychological peculiarities of individuals, the element of randomness in human behavior. And although this possibility cannot be merely dismissed, we cannot in good conscience adopt it, since it smacks of an ad hoc attempt to save the theory. We are not prepared to offer any definitive explanation of this anomaly, but we will point out that none of the observed cases concerned French chemists, so there is the possibility that the lack of personal contact was a factor.

Overall, the French data support the theory. While the British data often do not or are inconclusive, they do provide us with an example of what happens in a group resisting a revolution: in particular, the models of Chapter 5 showed how the phlogiston theorists reacted to the new paradigm and opposed it. All of the coefficients, to the extent that they were significant and with the exception of those relating the level of theory and quantification to paradigm choice, were in the same direction as those in the French data. The negative associations of quantification and level of theory with paradigm choice probably reflect the tendency of younger British authors who were beginning to shift to the new theory to avoid theoretical argument and weight relations, because of their deference to the leaders of the British chemical community.

On pages 26, 70, and 92 some question has been raised about the paradigmatic nature of phlogiston. Most historians assume that phlogiston theory was the guide for chemical research in the 1760s and 1770s. For example:

During the 1760s, before 'pneumatic' or gas chemistry emerged as an important subject, the prevailing explanation of combustion and calcination in Europe was one elaborated by Stahl and his followers. [Hufbauer, 1971, p. 119]

Since Lavoisier, like everyone else at the time [1772], adhered to the phlogiston theory... [Kohler, 1972, p. 351]

However, Perrin (1970), a Guerlac protege, appears to argue against this view. He notes that

> Like all young French chemists of his time, [Bucquet] had been trained in the phlogistic school of thought and had used its concepts and terminology in his two books published before 1774. [p. 136]

But he goes on to discuss "an undercurrent of doubt and controversy concerning [phlogiston]" (p. 142). He proposes three sources for this undercurrent: (1) a "residual scepticism," not among practicing chemists but among *physicists and mathematicians* and a few *"survivors* of 'pre-phlogiston' French chemistry"; (2) Meyer's theory of acidity and a "search for an accommodation"; and (3) "the discoveries of pneumatic chemistry" which "revealed *new facts* for which there were *no phlogistic explanations"* (pp. 142-143, italics mine). Using our theory of scientific revolutions, however, we can see that these observations merely confirm that phlogiston was a recently triumphant paradigm. While the three sources may have inspired Lavoisier or Bayen, they did not create serious problems for other phlogiston chemists. What Perrin sees as anomalies were not anomalies until Lavoisier made them so.

Point (1) is easily explained in terms of our theory: Physicists and mathematicians were not chemists and, not surprisingly, were not members of the paradigm group. Those who were practicing chemists were not threatened by this "undercurrent." Furthermore, there is no evidence that any of these mathematicians or physicists objected to the phlogiston theory until Lavoisier had already brought it into question. Finally, the "survivors" are clearly examples of diehards who refused to convert just as Priestley refused to convert to oxygen theory, and they too were left behind by the field.

As for source (2), I would argue that Meyer's theory was an example of puzzle-solving, that is, it consisted of a modification or adaptation of the basic paradigm to solve a particular problem. Perrin points out that "the two theories were not necessarily incompatible" (1970, p. 143).

Perrin's third point is a bit misleading. Of course there were new facts presented by pneumatic chemistry for which there were no *existing* phlogistic explanations — new facts arise for all new paradigms and lead to puzzle-solving, the essence of "normal" science. Footnote 9 on page 28 as well as the following observations show that phlogiston chemists were creating solutions to these puzzles:

> From the data provided by their experiments on gases, phlogistic pneumatic chemists inferred that distinguishable species of gases were only modifications of but one gas whose own properties were obscured and new properties gradually appeared as phlogiston accumulated in it. [Langer, 1972, p. 5]

> The first phlogistic theory of gases was embodied in the British pneumatic model established by Joseph Priestley in its most primitive form in 1772. [Langer, 1972, p. 47]

Our examination of the process of revolution shows, therefore, that Perrin's observations support the opinion that phogiston functioned as a paradigm rather than his view that it did not. This is also the conclusion of our data, which depict the long defense of phlogiston by its adherents.

The results of the hypothesis testing appear to generally support the ideas of Kuhn, for we were able to predict with much accuracy the process of change and to explain resistance in terms of the theory. Some modifications are in order, however. First, we have seen that there is some data, albeit inconclusive, indicating that paradigms do not always function as gestalts. However, when we take into account the fact that none of the exceptions was French and that the one we observed in Britain was not a prominent or professional chemist, we feel that this is still a useful heuristic assumption.

Second, the data show clearly that it was Lavoisier's new theory and his persistent — although at first hesitant — presentation of it from a position of strength in the Académie des Sciences that led to a crisis and the perception of anomaly by other chemists. This reverses the sequence set forth by Kuhn (although as we noted, he admits this possibility). There is good reason, furthermore, to think that this is the normal sequence: by Kuhn's own argument, scientists will not give up a paradigm until there is a candidate to take its place. Thus it is likely that anomalies will not become generally perceived — perception is the sine qua non for a communal crisis — until some maverick offers an alternative. Without this, we find precisely what occurred throughout the phlogistic camp: a series of modifications of theory in attempt to solve the anomalous results — and even then it is not clear that the results would have been seen as anomalous at all unless some scientist called attention to them. Priestley, for example, never recognized the existence of the anomalies of phlogiston and never converted. Clearly, other revolutions need to be examined to determine the usual course.

The advances over Kuhn's anecdotal description of revolutions consist in the details of the process which have been supported: the changing relations and levels of theory, quantity, and paradigm; the clear relation, often asserted but seldom examined, between age and early acceptance; the role of nationality and other subcommunities; the significance of the status and prestige of the originator or early converts; the great importance of personal contact and the actual viewing or participation in experiments by a convert.

With reference to Hagstrom's ideas, the information-recognition exchange model was found to be useful in directing our attention to journal publications and citations. There is the further suggestion that the presence of several career possibilities and the concrete manifestation of recognition in the French chemical community, such as election to the elite membership of the Académie or a government commission, helped to motivate the French chemists to perform.

Moreover, there is strong evidence that even recognition is not enough, that material support is necessary for the development of a viable scientific community. In the French case, the two factors are very difficult to separate, for the avenues to recognition were also the avenues to facilities and monetary support: election to the Académie led inevitably to a pension and often to a salaried post, as well as a laboratory; appointment to a teaching post in Paris, unlike the situation in Britain (or Germany until late in the century), also resulted in a salary, students, and facilities. Therefore, while some scientists may be motivated purely by the desire to advance knowledge, there is convincing evidence that communities of scientists do not flourish when that is the only motivating, or controlling, factor.

The evidence is also clear on the issue of social control. Chemists in Britain, where very little control existed — almost anyone could publish in the *Philosophical Transactions* — produced some very poor work, as noted by Trengove and by eighteenth century authors. Furthermore, the amateurism of Priestley, while it allowed him to do some very fine work, also permitted him to work on a variety of subjects, almost at random, rather than focusing his attention. This is not all bad, of course — Hahn even sees it as an advantage when he criticizes the Paris Académie for excluding certain theories, such as those of Lamark in biology (Hahn, 1971, p. 113). The data in our study indicate that once oxygen theory was accepted by Laviosier and his collaborators, it became the orthodoxy of the Académie.

So the information-recognition model has been shown to be a useful, but incomplete theory of the social structure and reward structure necessary for science. The additional factors of monetary rewards and working facilities are crucial — and unless one is independently wealthy (as were Cavendish and Priestley), it is difficult to subsist without these requisites.

Present-day concern in the scientific community about the pernicious effects of large federally-controlled grants (a concern which was more uniformly expressed during the period when the National Science Foundation was being established) and the directions in which they lead research reflect both the necessity for facilities and monetary rewards in science and the possibility of subversion of the scientific enterprise.

The differences between the organization of the Paris Académie and the Royal Society of London, as well as the more authoritarian orientation of French society in general, show further that science fluorishes in nondemocratic societies at least as well as in democratic (or egalitarian) ones. This casts doubt on the common notion that science and democracy go hand in hand, doubt which is strengthened by the evidence of the spurt in scientific growth in France under Napoleon, in Germany in the nineteenth century, and in the United States under the aegis of a huge, hierarchically organized federal bureaucracy.

Another aspect of Hagstrom's approach is the emphasis on communication and the relative lack of productivity by isolated chemists. Two findings support this point of view. One is the role of personal communication and contact alluded to above. The other may help explain the lack of productivity and/or resistance of British chemists — isolation. With only a small number of chemists scattered in several cities — London, Birmingham, and Edinburgh, for example — it was probably difficult in Britain to communicate or be professionally stimulated. Schofield's study of the Lunar Society (esp. 1963b) shows the importance of social contact: even though this group was primarily interested in practical, business concerns in the growing industrial city of Birmingham, they still produced some very good work and prominent chemists (Watt, Keir, Priestley).

Thus, this study of one of the most important eras in the history of chemistry has illuminated both the history and the sociology of the period. For the history of science it has added the important finding that great differences in the social structure of French and British chemistry led to previously unsuspected differences in the quality of chemical literature in the two countries. Furthermore, we have discovered a large gap in the ages of French and British chemists which contributed to British resistance to the Chemical Revolution. These results demonstrate the utility of quantitative methods in historical investigation.

For the sociology of science the study has supported the emphasis on non-logical factors in the process of scientific discovery and confirmation, indicating the importance of general social structure (the impact of the early rise of the Industrial Revolution in Britain), scientific organization (the hierarchical structure of the Paris Académie as opposed to the relatively democratic nature of the Royal Society), age, nationality, tradition (the emphasis on certain kinds of quantitative work in France but not in Britain), and personal contact in the process of revolutionary conversion. These results demonstrate the utility of focusing on the substance of science, in addition to its institutional structure.

For the philosophy of science the study has dealt another blow to the view that the process of confirmation of scientific theories is or can be amenable to purely deductive (or inductive) logic. The only possible conclusion is that there can be neither a neutral observation language nor an algorithm for determining which of two scientific theories is correct. In fact, the findings seem to raise serious doubts as to whether the term "correct" has a useful meaning in the case of science.

EPILOGUE

Given these discoveries, it is appropriate to do some speculation on the basis of the theory. In the beginning of the study I pointed out that we were not trying to explain the genius of one man, but rather to accept his work and focus on its reception by other scientists. However, I now think that we are in a position to explain why, based on the theory, Lavoisier was the scientist responsible for the Chemical Revolution.

We have discovered that the factors leading to acceptance of the oxygen theory included youth, emphasis on careful and essential use of the balance, and French nationality. Furthermore, we have observed the significance of

Académie membership and government salaries for the development of a professional approach to science, along with a tradition of journal publication. Finally, in Chapter 6 we noted the great importance of being a student of Rouelle.

These factors, several of which are derived from general theory and the rest discovered in testing that theory, show that Lavoisier was unique in his scientific structural position. He was part of the Continental tradition of weighing, he was a member of the Académie from 1768 and therefore had access to its facilities, and he was a student of Rouelle. These positions alone do not make him unique, however, for there were at least half a dozen men in this position. They may well have been as brilliant as Lavoisier; after all, there have been many brilliant scientists who have not been responsible for scientific revolutions. What does appear to make him unique, however, was his age. Lavoisier was the only person in the early 1770s who was the precise age to take advantage of the suggestive evidence, who could risk taking the bold step of putting forth a new theory. Of the two other men who questioned phlogiston theory at this time Turgot was not a scientist and Bayen was relatively old (47) in 1772. Of Rouelle's other students (who did not raise questions) Cadet was 41, Darcet 47, Baumé 44, Macquer 54, and while Sage was only 32 he was probably too peripherally involved with chemistry to recognize anomalies (he never did). Bucquet, on the other hand, was 26, perhaps too young − and we know that he was the first convert to the new theory before 1780, at which time he would have been 34 if he had lived.

Lavoisier was just 29 in the crucial year of 1772 − old enough to have studied with Rouelle, to have absorbed the details of the phlogiston theory, to have developed an interest in combustion, acids, and weighing, and to have been elected to the most renowned scientific organization of his day. Yet he was also young enough to take advantage of his situation. There is also evidence that he fit the model of the recognition seeker quite well: he often rushed into print rapidly, at least once with incorrect experimental results he had stolen from Priestley. Thus, there is a strong prima facie case that anyone who combined these characteristics with intelligence could have created a revolution just as Lavoisier did.

One final factor I wish to emphasize is that Lavoisier was in fact a good phlogiston chemist before doubts began to nag him, as were Morveau, Berthollet, and Fourcroy. Although it is easy for a person outside the field, such as Turgot or Bayen (an apothecary for the army), to raise questions or readily accept another's new theory, it is difficult for an outsider to create the new theory himself. Thus, it is particularly significant that Lavoisier excelled in phlogiston chemistry and that he had studied with Rouelle. At the same time, of course, he had not been involved in phlogiston chemistry long enough to have become inflexible: he knew enough to recognize problems, he believed that weight was important and took notice of others' discoveries that metals gained weight when calcined, and he was young and flexible enough to question the explanations others offered. And perhaps he was also arrogant and ambitious enough to take a risk and "enter the lists against [Stahl]."

Appendix A:
Methodology and Data

Assuming that social control in science operates through a communication system in which research is exchanged for recognition, we will focus on formal communication, i.e., publication in periodical literature. In order to make the theory concrete and testable, therefore, the operational versions of the hypotheses are stated in terms of journal articles.

Publication does not occur randomly, and patterns develop over time. This notion of pattern may be somewhat loose, but it is clarified by the text. The fact that a revolution or paradigm shift in science involves the replacement of old theories and practices by new ones implies that there will be changes in the communication patterns, that is, in the characteristics of journal articles.

The Chemical Revolution involved a shift from a paradigm based primarily on phlogiston as the principle of combustion and calcination to one based on oxygen. The shift began shortly after 1772 in France and was completed before 1800, except for a few diehards. Our investigation begins in 1760, in order to establish some sort of base for normal science, and ends in 1795, when science in France was in great turmoil from the French Revolution.

The data consist primarily of information characterizing all or nearly all chemical articles written by French or British authors published in France or Britain (including Ireland and Scotland) from 1760 to 1795.[1] I obtained this information by locating and examining all periodicals available and reading briefly all of the articles; these were then characterized and their citations recorded.

Bolton (1893) and Scudder (1965) provided a list of all the journals which presumably would have any chemical information. In addition to chemical and general scientific journals, the list included journals devoted primarily to medicine,

[1] For comparison and completeness, German periodicals and a small selection of German articles were also checked.

pharmacy, mineralogy, as well as those of a general learned nature. The 322 periodical titles[2] thought to contain chemical information consisted of 32 British, 48 French, and 242 German, of which 244 were located in the *Union List of Serials* (1950) or in card catalogues of the libraries visited: 31 British, 41 French, and 172 German. Some titles, particularly the German ones, were not available either in libraries I was able to visit[3] or on interlibrary loan (they are too rare to be lent), so I examined only 216: 31 British, 40 French, and 145 German. I was, therefore, able to get a sample of 97% of the British, 83% of the French, and 60% of the German titles. This is undoubtedly a conservative estimate of the total percentage of journals containing chemical information which were examined, because some of the missing titles are probably duplicates[4] of those found and others are, no doubt, devoid of chemical information. In any case, I assume that all of the important French and British journals were examined.[5]

Eighty-six of the journals contained no chemical information: 13 French, 16 British, and 57 German. Consolidation of continuations under different titles gave a final sample of 118 journals (23 French, 15 British, and 80 German).

Of the 38 French and British journals, 12 French and 8 British contained original chemical articles.[6] I selected 724 French and 134 British articles from

---

[2]I use the word "titles" advisedly, as this number includes various name changes that several of the journals underwent, for example, *Introduction aux Observations sur la Physique . . .* to *Observations sur la Physique . . .* to *Journal de Physique . . . .*

[3]This required visiting twenty libraries across the country as these materials are normally too old to be lent on interlibrary loan. In some cases they are difficult to locate even in the library that has them. Some additional titles were discovered in some of the libraries, either in the stacks or the card catalogues. Some of them were in Bolton but not in the *Union List* and some were additions to Bolton.

The twenty libraries were: Academy of Natural Sciences Library, Philadelphia; American Museum of Natural History Library, New York; American Philosophical Society Library, Philadelphia; Boston Athenaeum; Boston Public; College of Physicians Library, Philadelphia; Countway, Boston; Engineering Society Library, New York; Library of Congress; National Library of Medicine, Bethesda, Maryland; New York Academy of Medicine Library; New York Public; and the libraries of the following Universities: California at Berkeley, Columbia, Harvard, Johns Hopkins, Massachusetts Institute of Technology, New York, Princeton, and Stanford.

[4]That is, they are either alternate titles or continuations under other names. Evidence of this was found in several libraries: the journals were listed by one name in the *Union List* and another in the card catalogue, although the titles used in the *Union List* are, supposedly, those submitted by the libraries holding the periodicals.

[5]Evidence for this is that no French or British journals were cited in the articles examined which were not on the list, and the only journals I can find mentioned in historical literature that might be important but which were not examined are *l'Avant Coureur* and *Journal Polytype*. *L'Avant Coureur* appears to have published news and summaries rather than original articles (Smeaton, 1959) and it is unlikely that *Journal Polytype* contained enough articles (if any at all) to modify any of the findings in this study.

[6]Twenty-five of the German journals also contained original articles.

these 20 journals, presumably all of the major papers and almost all of the minor ones which were first published in periodicals. Selection presented some problems in cases of borderline articles in periodicals that were not primarily chemical. These articles were in such fields as medicine, pharmacy, mineralogy, or natural philosophy, all of which at this time had strong ties with chemistry. I tried to include all articles which appeared to present facts of a chemical nature, as opposed to those which were purely medical, mineralogical, or physical. In difficult cases I tried to err in the direction of inclusion. This may introduce some extraneous (nonchemical) variance, creating problems for the hypotheses, but it should not be serious. Topics which one might treat as chemical, but which I did omit, were the structure of crystals (physical), electrical phenomena, and such physical topics as heat.[7] I also omitted papers that were reports of committees, and which usually summarized results in some subfield or on some particular substance but did not present original research. Some original results may have been missed in these cases, but the omission is very minor, for there were very few of these papers.

THE VARIABLES

I characterized each article in terms of ten variables and the author's name(s). Multiple authorship, ranging from two to four, was noted: all such articles were French and their were only 48 of them, about 6% of the French sample. The ten variables are as follows:

1. Nationality of the author of the article: French or British.

2. Age of the author of the article at the time of publication, ranging from 18 to 70. For some purposes, this variable was dichotomized at age 40.

3. The journal in which the article was published for the *first* time, ignoring reprints, revised versions, and extracts from previously published articles.[8]

4. Date: the year in which the article was published, ranging from 1760 to 1795. Because this is a study of *formal* communication and publication, I did not use years of submission to journals or of presentation to societies, but only the actual date of publication of a journal, including memoirs of societies. This

---

[7]This last category, which would today be called physics, or perhaps physical chemistry, is probably the most serious omission. There is evidence that "physical" considerations, as opposed to "chemical" considerations, were important in Lavoisier's thinking (especially Morris, 1972, and Siegfried, 1972) and also in the work of British chemists (see McCormmach, 1969; Schofield, 1967), but it is doubtful that the few such articles would seriously modify any of the findings.

[8]We may, therefore, have omitted a few articles which had "extract" in their title (although they had not been published previously) or included a few articles that had been published previously but that did not indicate this fact. We doubt that there were many in either category.

may result in some discrepancies between the dates I report for some events and those most historians give. The discrepancy due to the delay between presentation and publication is most severe in the cases of the Académie Royale des Sciences in Paris and the Société Royale de Médicine in Paris. In some cases articles were revised between the date of presentation and the date of publication. Further- more, some papers which appear in a volume of a society for a particular year (the volumes were dated by year of presentation) were actually presented in a subsequent year. For these reasons, as well as for the sake of simplicity and consistency, I have uniformly used the date of publication.[9]

For clarity of presentation, I have generally collapsed time into nine (or fewer) categories of varying length corresponding to significant historical events. The first two intervals, 1760-65 and 1766-71, comprise the base period of normal science, with 1766-71 the period in which pneumatic chemistry was developing in Britain. The next interval, 1772-77, is the period during which Lavoisier was developing his new theory: 1772 is the year he deposited the sealed note with the Académie and 1777 is the year of his first frontal attack on phlogiston. The next two intervals, 1778-1780 and 1781-84, are the periods during which Lavoisier was trying to justify his theory to others and meet the objections of his opponents. They extend to the year in which he conducted his crucial experiment (the decomposition and recomposition of water) which led to the rapid conversion of the French chemical community. The period is divided into two arbitrary intervals. The next three intervals are as follows: 1785-86, starting with the year of Lavoisier's crucial experiment; 1787-88, starting with the year of publication of the new nomenclature; and 1789-90, starting with the year of publication of Lavoisier's *Traité Élémentaire* and the founding of the *Annales de Chimie*. Taken together, these trhee periods form the years of debate and conver- sion, in which the fate of the new theory was decided by the French chemical com- munity, and in which French chemists had to risk their careers and decide. The final period, 1791-95, is viewed as the consolidation phase, a return to a new period of normal science.[10]

5. The length of the article in pages. This ranges from reports of one to two pages to long treatises of more than 50 pages. The longest ones were usually on

[9]This practice, which seems to me the best of several strategies, may result in some problems. Most notably, responses to some particular discovery could be spread over several years somewhat artificially by one scientist responding in a journal which was published frequently and rapidly (such as *Observations sur la Physique* . . .) while another (or even the same one) responded in a journal which published either infrequently or after some delay (such as the *Mémoires* of the Paris Académie). Although I have no evidence that this sort of thing seriously affected the results, it does introduce at least potential measurement error into this variable.

[10]Since the Terror of the French Revolution had such a devastating effect on publication and community life, I have divided this period into two, 1791-92 and 1793-95, in the data for productivity in Chapter 4, but nowhere else, as there was no evidence that any relation- ships were noticeably affected.

practical or applied topics and were often written in competition for a prize offered by a scientific society.[11]

6. Paradigm choice, ranging from support of phlogiston to support of oxygen. The basic scale consisted of seven categories: (a) overt support of phlogiston; (b) use of the old nomenclature after 1787; (c) expressed neutrality; (d) no information with respect to phlogiston or oxygen; (e) use of the new nomenclature; (f) expressed antiphlogistic ideas; (g) overt mention of oxygen.

The categories are not mutually exclusive, but the following conventions made them so: category (a) takes precedence over (b), that is, use of old nomenclature including the favorable mention of phlogiston results in the article being placed in category (a); category (g) takes precedence over (e) or (f) in the same way; category (e) takes precedence over cateogry (f). Most of the analyses used a three-cateogry scale derived from this one. Categories (a) and (b) became "phlogiston." However, for Britain there were exceptions, for (b) includes at least one article using the old nomenclature in 1791 while attacking phlogiston theory. For such cases, though they were extremely rare, the articles were, of course, categorized in terms of the paradigm they appeared to support. Category (c) remained "neutral." Categories (e), (f), and (g) were collapsed into "oxygen," while category (d), considered missing data, resulted in the article being eliminated from some analyses. Table A.1 which shows the distributions of articles on the two scales, should make the relationship clearer.

Therefore, "phlogiston" or "uses the phlogiston paradigm" is, technically, an abbreviated form of "uses the phlogiston paradigm or the old nomenclature after 1787," and "oxygen" or "uses the oxygen paradigm" abbreviates "uses the oxygen paradigm or the new nomenclature or is antiphlogiston." I assumed that use of the new nomenclature (or lack of such use) implies use of the oxygen paradigm (or support of the phlogiston paradigm), but being antiphlogiston is not quite the same as using the oxygen paradigm. I occasionally make use of these finer distinctions.

7. Level of theory, ranging from purely descriptive to purely theoretical. This scale consists of eight categories: (a) purely descriptive, no explanations; (b) primarily descriptive, very little explanation or explanations restricted to non-chemical results; (c) a moderate amount of explanation, many results left unexplained; (d) relatively complete explanations, but no theoretical ideas used; (e) explanations using a little theory; (f) explanations using a moderate amount of theory; (g) a major theoretical component in the explanations, but the theory is used to account for the results rather than the experiments used to explicate the theory; and (h) purely theoretical, the purpose of which is to attack

[11]These prize competitions, usually offered for solving practical problems requiring scientific knowledge, sometimes resulted in important papers. A study of these prizes and their effects on science, scientific societies, and society should prove interesting and informative.

TABLE A.1
Distribution of Articles on the
Two Paradigm Choice Scales

	Phlogiston	Neutral	Oxygen
French			
Phlogiston	208	0	0
Old nomenclature	11	0	0
Neutral	0	5	0
New nomenclature	0	0	80
Antiphlogiston	0	0	24
Oxygen	0	0	94
"No information" = 302			
British			
Phlogiston	49	0	0
Old nomenclature	7	5	1
Neutral	0	1	0
New nomenclature	0	0	5
Antiphlogiston	0	0	1
Oxygen	0	0	2
"No information" = 63			

or defend a theory with experimental results presented only to demonstrate and prove the theory.

This is the most doubtful scale, and it suffers from several defects: it is subjective, relying on judgments of what is a "little," a "moderate amount," or a "complete" accounting for results; the criteria of placement may have varied somewhat from year to year; it may not be a unidimensional measure of increasing level of theory, although it is treated as one. Perhaps the most serious difficulty is that the categories are not necessarily mutually exclusive, reflecting the underlying lack of unidimensionality. For example, an article could contain very little explanation, meaning it should be put in category (b), but the explanation given might be highly theoretical, meaning it should be in some higher category. In practice, I have forced the categories to be mutually exclusive and have placed the few articles similar to the example just described one category higher than their basic characteristic would indicate.

For some of the presentations these eight categories are collapsed into three: "descriptive" (a and b), "mixed" (c, d, e, f), and "theoretical" (g and h); or just two: "nontheoretical" (a, b, c, and d) and "theoretical" (e, f, g, and h). The latter division is based on the reasoning that the former group contains essentially no mention of theory while the latter contains only articles that have theoretical aspects; so it is, perhaps, the surest measure.

A problem even with this latter division is that what might be considered a theoretical or even hypothetical idea at one point in time may be considered a mere fact at another time; for example, the identification of a particular substance as "oxygen." This last type of problem is inherent in any study of a scientific revolution and is related to the problem Kuhn discusses regarding attempts to date the discovery of oxygen (1962, pp. 53-56).

8. Level of quantification: whether the article was qualitative or quantitative. An article was considered quantitative if the author measured both the quantity of materials entering into a reaction and the quantity of materials resulting from the reaction. Unfortunately, the conception of what constituted a quantitative article varied somewhat over the years that this study was carried out. Some were not considered quantitative unless a substantial portion of their results were quantitative, while others were considered quantitative if only one quantitative analysis appeared in the paper. I am sure that the "quantitative" group is more quantitative than the "qualitative" group, but of course I cannot say that about individual articles. However, since there is no reason to believe that the errors are not random, the effect is to reduce correlations, making the hypotheses harder to prove.

There is a further problem which may affect the relationships between this variable and the others and, therefore, the hypotheses. That is the problem of pseudo-quantification or "recipes" mentioned in Chapter 2.[12] If one measures the amounts of substances in order to follow a recipe to get a certain amount of some product, rather than measuring to find out exactly how much of the reagents is necessary to produce an exact amount of a product, then the mental process is quite different from that accounted for by the theory. Such recipes had existed in chemistry (and alchemy) for a long time and did not play the same role as the essential quantification of Lavoisier or Guyton when measuring the weights involved in a reaction such as combustion. Since the hypotheses concerning the relationships of time, level of theory, and paradigm choice to quantification, were based on the assumption of true quantification, the presence of articles in the quantitative category which were, in fact, not quantitative in this sense may affect the results in a conservative manner (that is, they may depress the relationships and make the hypotheses harder to demonstrate). Therefore, the text mentions this problem where it might have had an effect.

9. Affinity: whether or not the article used affinities, a major theoretical and practical tool.

10. Citations: the names of authors cited and the number of authors cited in each article, including only citations to chemical works and authors. I had originally hoped to include the actual books and papers cited, but due to the style of citation during this period, this was impossible — such references appear only a small fraction of the time. This information was also used in the form of the number of *distinct* authors cited in each article, which ranged from 0 to 24 and was scaled as 0, 1, 2, 3, 4, 5, 6, 7, 8, and 9 or more.

[12]See, for example, the characterizations of Black's work by Multhauf (1962) and Guerlac (1961a).

In addition to these characteristics of each article, I used some biographical information about as many of the authors as I could find in standard sources (see the discussion of the samples).

THE SAMPLES

The basic sample consists of all 858 articles noted above: 724 French by 159 authors and 134 British by 48 authors. I make little use of this sample in the presentation, although almost all of the analyses were checked against it to see if there were any important differences, noted in text. The theory is based on the assumption that one is dealing with a community of scientists. In order to be a member of such a community, one must be recognized. Although mere publication is a form of recognition, I do not feel that such recognition implied that one was a functioning member of a scientific community in the eighteenth century. In view of the fact that the total pool of authors was small (about 200) and that there was essentially no refereeing,[13] citation seems to be a reasonable minimal criterion of membership in the chemical community. Therefore, for most of the analyses I have eliminated all papers of authors who were never cited by other authors in the basic sample, omitting 108 French articles by 66 authors and 33 British articles by 21 authors.[14]

The authors eliminated were, in several senses, minor. First, they were not important enough to have been cited by any other chemist.[15] Second, few of them published more than one article (thus contributing little for which to be recognized): among the French only five published more than three,[16] 14 published two or three,[17] and 47 published only one or a part (coauthor); among the British, seven published two or three,[18] and 14 published one.

[13] One could argue that the *Mémoires* of the Paris Académie were refereed and that anyone who published there was a member of the community. Clearly, publication there implies recognition, since it means membership in that elite group. As a matter of fact, this criterion excluded only one member (Fougeroux de Bondaroy) who published any chemical articles, and he was not primarily a chemist.

[14] Because such elimination might have affected some of the results, some of the analyses were also done for the basic sample. None of the important conclusions, however, was affected.

[15] Of course, some of them may just have been too new to the field to have received citations by the time of our cutoff date. However, even though such authors may have become important later, they were not so during the period under study.

[16] These were de Lunel – 7, Berniard – 6, de la Planche – 5, Opoix – 4, and Fougeroux – 4.

[17] These were Legendre, Mollerat de Souhey, Puymarin, fils, Macquart, Couret, Bosc d'Antic, Marges, Torchet, Fourcy, Pasumot, Ribaucourt, Planchon, Mesaize, and Bouvier.

[18] These were Monro, Fordyce, Willis, Wall, Morris, Clegg, and Austin.

TABLE A.2
Distribution of "No Information" Articles
by Level of Theory Controlling for Country
(Cited Sample)

	Descriptive	Mixed	Theoretical	Total
		French		
Percentage				
"no information"	59	28	7	41
Number of cases	(303)	(243)	(70)	(616)
		Kendall's tau-b = .37, p <.001		
		British		
Percentage				
"no information"	54	45	5	41
Number of cases	(39	(42)	(20)	(101)
		Kendall's tau-b = .30, p <.001		

Finally, most of them were not found in any of the standard bibliographic sources.[19] Although the lack of biographical information makes it impossible to check, I feel those omitted were likely to be very young (and were published too late to be cited), to come from outlying areas, and to practice primarily in fields peripheral to chemistry.[20] They were of little importance for chemistry.

The sample which is left, consisting of 616 articles by 102 French authors[21] and 101 articles by 27 British authors, is called the "cited" sample.

In order to get a sample appropriate for the more complex analyses, including the causal models, I made two further adjustments. First, because the theory involves paradigm choice, all articles which did not make a paradigm choice or express clear neutrality, (the articles classified "no information") had to be considered missing data. Second, because age is an important variable, I also eliminated all articles for which the author's birthdate could not be found. The

[19]The standard sources consulted were Partington (1962), *Dictionary of National Biography,* Poggendorff (1965), Michaud and Michaud (1811) and the *Index Biographique* of the Paris Academy. Taton (1964), also proved very useful. In addition, I consulted the work of Cartwright(1967), Coleby (1952, 1953, 1954), Gibbs (1952a), Gibbs & Smeaton (1961), Mayer (1970), Rappaport (1960), Russell-Wood (1951), Sivin (1962), Smeaton (1962, 1967), and Smith & Moilliet (1962).

[20]Some evidence of this is provided by the journals in which they were published, which were minor or peripheral – see Table A.5.

[21]Of these 102 authors, nine were not cited but were coauthors with authors who were cited.

TABLE A.3
Distribution of "No Information" Articles
by Quantitativeness Controlling for Country
(Cited Sample)

	French		British	
	Qual.	Quant.	Qual.	Quant.
Percentage "no information"	46	29	42	38
Number of cases	(412)	(204)	(69)	(32)
	Tau-b = .16 $p < .001$		Tau-b = .04	

resulting sample, consisting of 351 French articles by 57 authors and 58 British articles by 16 authors, is the analytic sample.[22]

The major differences between the cited and analytic samples are due primarily to the "no information" criterion because of the close relationship of this category and level of theory, shown in Table A.2. As would be expected, descriptive articles (those which do not explain their results) often provide no clue as to which paradigm the author may support.

The French "no information" articles were also disproportionately qualitative as Table A.3 shows, although the British were not.

Use of the theoretically motivated "no information" criterion eliminated 250 French and 41 British articles, while use of the accidental age criterion elimnated only an additional 15 French and 2 British articles.[23]

So that we can better assess the effects of the selection process, Table A.4, which summarizes the values of the major variables,[24] should make clearer the differences between samples.

[22]The underlying justification for this, and the previous, paring-down of the sample is taken from Tukey and Wilk (1970, p. 376):

> Using human judgment in the selection of parts of the data for analysis, or in cleaning up the data by partial or complete suppression of apparently aberrant values is natural, sensible, and essential. Data is often dirty. Unless the dirt is either removed or decolorized, it can hide much that we would like to learn.

[23]As might be expected, most of the authors for whom I could not locate a birthdate were not cited (they are clearly less important men than those who remain) and, therefore, had already been eliminated. I was able, in fact, to find birthdates for the authors of 575 of the 616 French articles and 99 of the 101 British articles in the "cited" sample.

[24]The complete distributions of all the variables in the three samples used and in the noncited group may be obtained from the author.

TABLE A.4
Values of Selected Variables by Sample

Variables	Basic	Cited	Analytic	Noncited
		French		
Date	82.3	82.7	84.4	80.1
Level of theory	3.03	3.11	3.91	2.55
Quantitative	(31)[a]	(33)	(39)	(18)
Oxygen/phlogiston	1.95	2.00	2.02	1.61
Use affinity	(16)	(16)	(22)	(13)
Pages	8.6	8.6	9.6	8.8
Citations	3.27	3.38	3.90	2.64
Number of articles	724	616	351	108
Number of authors	159	102	57	66
		British		
Date	82.7	82.8	85.8	82.1
Level of theory	3.46	3.83	4.66	2.30
Quantitative	(31)	(32)	(35)	(27)
Oxygen/phlogiston	1.34	1.28	1.22	1.64
Use affinity	(22)	(19)	(16)	(33)
Pages	10.9	12.0	11.3	9.8
Citations	3.22	3.56	4.10	2.15
Number of articles	134	101	58	33
Number of authors	48	27	16	22

[a] ( ) = Percentages. Numbers of pages are medians. Other values are means.

There are some notable differences between the values for the two countries, most of which are discussed in detail throughout the study. These include the great differences in the numbers of articles and in the values of level of theory and oxygen/phlogiston. The only other difference which is significant (statistically, as well as substantively) is that the French wrote somewhat shorter articles. This is, in part, due to the fast pace of French chemical activity during the period under study, reflected in the large number of articles that French chemists published as they strove for relatively quick dissemination of results — this is especially true of Lavoisier (Guerlac, 1961b). The difference is also partly a reflection of the work of Sage, a mineralogist who wrote 61 articles, more than anyone else, of which a large fraction were short (1-3 pages).

In general, each stage of selection produces a higher quality sample, in the sense that the average values of the basic variables (time, oxygen, theoretical level, and level of quantification) are increased (except for the value of oxygen among British papers, which declines to the phlogiston end of the scale). The criteria, as intended, produce a more professional sample of papers — those that were the most involved in the chemical world of the late eighteenth century.

A further example of this professionalization is the decline in the proportion of articles classified "no information." For French articles, this proportion ranges from 48 for noncited, to 42 for basic to 41 for cited (and 0 by definition for analytic). Among British papers it ranges from 67 for noncited to 47 for basic to 41 for cited.

Among the most interesting of the differences among samples are those involving the journals in which the articles were published. I stated that the noncited authors published in journals of lesser prestige or importance, either in ones peripheral to the field or to the geographical center of the community. The data in Table A.5 support this conjecture: each stage of selection increases the representation of the major journals at the expense of the more peripheral (both in content and geography) journals, as shown by the progressive increase in the proportion of articles in the top two or three journals, and the progressive decrease in the proportion in applied, medical, and outlying journals.

TABLE A.5
Percentage of Articles in
Kinds of Journals by Sample

Journal Types	Basic	Cited	Analytic	Noncited
French				
Top three[a]	71.7	76.5	85.2	44.5
Applied[b]	15.7	11.7	6.6	39.0
Medical[c]	14.2	10.4	4.3	36.2
Outlying[d]	6.3	6.0	4.3	6.5
Savans Etrangers	6.4	5.7	4.0	10.2
British				
Top two[e]	78.4	86.2	87.9	54.5
Applied[f]	7.4	2.0	1.7	24.2
Medical[g]	2.9	1.0	0.0	9.0
Outlying[h]	13.4	11.0	8.6	21.2

[a]*Observations sur la Physique. . ., Mémoires* of the Paris Académie, and *Annales de Chimie.*

[b]*Journal de Médecine, Médecine Eclairée. . . , Journal des Mines, Mémoires,* of the Société Royale de Médecine, Paris, and *Mémoires* of the Société d'Agriculture, Paris.

[c]*Journal de Médecine, Médecine Eclairée. . . , and Mémoires* of the Société Royale de Médecine.

[d]*Philosophical Transactions* and *the Mémoires* of the Academies at Dijon, Montpellier, and Toulouse.

[e]*Philosophical Transactions* and *Observations sur la Physique.*

[f]*Transactions of the Royal Society of Arts, Medical Museum, New London Medical Journal,* and the *Transactions of the Royal College of Physicians.*

[g]*Medical Museum, New London Medical Journal,* and *Transactions of the Royal College of Physicians.*

[h]*Memoirs* or *Transactions* of the Royal Societies at Manchester, Edinburgh, and Dublin.

Some journals I have eliminated altogether: among French journals, the *Mémoires* of the Académie at Monpellier in the analytic sample, while among British journals, when I omit noncited authors, the *Medical Museum* and the *New London Medical Journal* have been eliminated and, in addition, the *Transactions* of the Royal Society of Edinburgh and of the Royal College of Physicians are eliminated in the analytic sample.

The difference between countries is also interesting. The French published a relatively greater proportion of their articles in applied journals (including medical ones) than the British, which may help to account for the greater proportion of theoretical articles in the British sample (see Chapter 4 for further discussion of this point). On the other hand, the British published somewhat more of their articles in outlying journals: it would appear that there was more centralization in French scientific organization. The difference is even more distinct when we eliminate British articles published in the (Parisian) *Observations sur la Physique* ... and *Annales de Chimie:* the proportion of outliers then becomes 16.2, 14.1, and 11.6 in the basic, cited, and analytic samples, respectively.

TECHNIQUES OF ANALYSIS

I have used three different kinds of units for the analyses: journals, authors, and articles. The journals are used as units only briefly in Chapter 3. The authors are used primarily in Chapter 3 as producers of articles and in Chapter 6 as citation receivers. For most of the discussions, however, articles are the units of analysis. It is important to keep these distinctions in mind, for they affected certain decisions as to how to treat the data. For example, for the causal models I use a sample in which all "no information" articles have been eliminated (the analytic sample). Where the distinction is particularly important I have made note of it in the text.

Two basic techniques were employed to test the set of hypotheses: cross-tabulation and path analysis (see Duncan, 1966; Land, 1969; Wright, 1921, 1934, 1954). Cross-tabulations are used to look at the basic distribution of data and to test most of the hypotheses in a crude way. The strength of cross-tabulation lies in its graphic display of data: one can see how the data are distributed among the categories and get some feel for the patterns in these distributions. However, it is quite difficult to look at more than a few variables at a time, since there are a great number of categories to look at, and the numbers in each of the categories become very small, making them difficult to follow. So for clarity in presentation and understanding, I make the assumptions necessary for multiple regression and use path analysis to examine the more complex relationships.

The assumptions of linearity and additivity are made initially and discussed briefly in the text, after which they are relaxed for a more sophisticated consideration of the models. The assumption of uncorrelated errors is made and seems no more problematic than in most other research. The assumptions of normality and homoscedasticity, necessary for tests of significance (Blalock, 1972, p. 367), are of little concern for the French because of their reasonably large number of cases. Since, in a sense we do have a universe — if we see the population to be (cited) French and British chemists — significance may not be an issue. On the other hand, because we wish to generalize to the population of all revolutions, we do not have a universe or even a random sample. Therefore, I make the assumptions, report significance levels to give some indication of stability (of at least the signs of the coefficients), and indicate in the text where caution is necessary.

The assumption of path analysis that the variables are measured at the level of an interval scale is problematic for some of our measures. Of those in the path model, time and age are interval scales, while level of quantification and nationality are dichotomies, which may be treated as interval scales: an interval scale being, essentially, an *equal-interval* scale, and because there is only one interval in a dichotomy, it is equal (to itself). Of the two remaining variables, level of theory and paradigm choice (oxygen/phlogiston), the latter has only three categories, with almost no cases in the middle one, allowing us to consider it a dichotomy. This leaves only level of theory to consider. I *assume* it to be an interval scale variable for the purposes of the analysis. Furthermore, there is reason to believe that a relaxation of this assumption is not serious: Boyle (1970), Bohrnstedt and Carter (1971), and others have shown that this assumption can be relaxed without much danger of distorting the results. In view of this, the advantages of causal modeling (by path analysis) outweigh the possibilities of misinterpretation due to the weakness of measurement. Furthermore, in general it is important to push measures to their limit, or even beyond, in order to take advantage of sophisticated techniques and to stress the need for *developing better measures,* measures adequate for our theories.

Appendix B:
Character and Contents of
Chemical Journals

The journals of the late eighteenth century which contained useful chemical information may be categorized in several ways. I have used a typology based on the information contained in a journal,[1] arranging journals in five categories: (1) journals containing original chemical articles; (2) journals containing reprints and/or translations of articles from other journals; (3) journals containing summaries of books, articles, and/or other journals; (4) journals containing news of science (meetings, discoveries, prizes, etc.) and reviews of various publications (these normally occur together); and (5) journals containing letters (correspondence between scientists not classified as articles).[2] These categories are not mutually exclusive: many journals contain more than one kind of information.

The typology and distribution of journals is displayed in Table B.1. The numbers are not totally accurate. I have offered conservative estimates of the journals in each category and have generally tried to place journals in the most appropriate category (i.e., dominant orientation), except where more than one categorization is clearly appropriate.

Of the 118 useful journals, 19% (23) were French, 13% (15) were British, and 68% (80) were German.[3] Of the 45 journals containing original articles,

[1]See Garrison (1934) for another useful typology. This article is also helpful in tracing title changes over time.

[2]As noted in Chapter 2, many letters were treated as articles because they were indistinguishable in content from short articles — refereeing was not a factor in the period under study.

[3]Compare this with Garrison's results for medical and scientific periodicals for the entire eighteenth century (1934, p. 300): 401 (73%) German, 96 (18%) French, and 50 (9%) British (not counting those of other countries); or for just scientific societies and scientific periodicals (his categories XII and XVII): 115 (66%) German, 40 (23%) French, and 20 (11%) British.

TABLE B.1
Numbers of Periodicals Containing
Various Types of Information by Country[a]

	France	Great Britain	Germany	Total
Original articles	12	8	25	45
Reprints, translations	8	3	28	39
Summaries	7	3	19	29
News and reviews	7	4	29	40
Letters	3	0	3	6
Total	23	15	80	118

[a]Numbers in the table sum across but not down since the categories are not mutually exclusive.

27% (12) were French, 18% (8) were British, and 56% (25) were German.[4] It is clear that the German publications overwhelmingly dominated every category except "letters." However, their relative percentage of journals containing original articles was somewhat less than their percentage of all journals, probably because German journals of all categories tended to reprint, translate, and summarize other publications rather extensively. One gets the impression that the Germans had the potential, at least, of being the best informed scientists and intellectuals of the period.

The high numbers for Germany, however, may be partly due to the non-unification of the German city-states in the eighteenth century,[5] resulting in repeat printing from city to city, a phenomenon unlikely in France. Furthermore, the numbers for France and Germany — and to a lesser extent Great Britain — are also inflated by the tendency for journals to have been published for short periods of time. Thus the numbers for any short time period are significantly lower than the totals, as shown in Table B.2. This table also indicates that the journals did not exhibit the same magnitude of growth as authors and articles. Aside from the period 1760-65, in which there were relatively fewer journals, there was marked growth among *all* journals only in Britain (from 3 to 11) and Germany (from 18 to 33), much of it after 1790.[6]

[4]The other numbers in the table are approximate, so no percentages are given.

[5]See Hufbauer (1971, Chapter 4) for a similar explanation.

[6]Similar results for journals containing original articles are presented and discussed in Chapter 3.

TABLE B.2
Numbers of Journals over Time by Country

	1760 -65	1766 -71	1772 -77	1778 -80	1781 -84	1785 -86	1787 -88	1789 -90	1791 -95
					All journals				
French	8	10	12	11	11	11	10	10	12
British	3	3	3	3	5	6	7	6	11
German	14	18	24	25	31	26	29	27	33
Total	25	31	39	39	47	43	46	43	56

DESCRIPTION OF PARTICULAR JOURNALS

In this section I will characterize briefly the major journals in terms of the primary variables: nationality of authors, years of publication, paradigm choice, theoretical level, quantitativeness, and age.[7] Lesser journals, when they are of interest, are mentioned only in passing.

Distributions on the following measures provide a profile of each journal: percentage "oxygen" includes the categories "antiphlogiston," "oxygen," and "new nomenclature" of our basic paradigm choice scale, while percentage "phlogiston" includes "phlogiston" and "old nomenclature" (after 1787); percentage "theoretical" refers to the highest category of our three-category scale of theoretical level; and the percentage of authors under 40 is accompanied by the number of articles, for we do not have ages for all the authors in the cited sample.[8]

1. *Observations sur la Physique, sur l'Histoire Naturelle, et sur les Arts,* started as *Introduction aux Observations, . . .* and continued as *Journal de Physique, de Chimie, . . .,* also called Rozier's Journal as it was started by l'Abbé Rozier in 1771. Abbreviated herein *Obs. Phys.*

This was the major French scientific periodical of the period.[9] It also published the second largest number of British articles, as well as those of other countries, including Germany, Sweden, and Italy. It was begun primarily to keep up with

[7]This is for the cited sample. The numbers of articles per journal in other samples may be obtained from the author.

[8]The overall figures for some variables may be misleading for those journals that did not publish throughout the entire period under study. Following the profiles, therefore, we give comparisons over time of the important variables for major journals.

[9]As such, it has received a great deal of attention by historians, including McKie (1958) and Neave (1950, 1951a, b, c, 1952).

the outpouring of papers in pneumatic chemistry at the beginning of the 1770s.[10] One of its advantages over many other journals, particularly those of scientific societies, was the rapidity of publishing – a paper could often appear in print less than a month after submission.

Of the 196 French articles published on *Obs. Phys.* 12% supported oxygen and 41% supported phlogiston, 15% were theoretical, 28% were quantitative, 16% used affinities, and 42% were written by men 40 or younger.

Of the 22 British articles published in *Obs. Phys.* 0% supported oxygen and 64% phlogiston, 27% were theoretical, 27% were quantitative, 32% used affinities, and 9% were written by authors 40 or younger. With the exception of the extremely high percentage using affinities, these percentages fit the British pattern more closely than French (compare the distributions for *Philosophical Transactions* and those in Chapter 4).

2. *Histoire et Mémoires,* Académie des Sciences, Paris, began in 1666 and was interrupted by the French Revolution and the subsequent reorganization of the Académie: the last volume before the reorganization of 1795 was published in 1791. Abbreviated herein *Méms., Paris.*

This was the premier publication of the major scientific institution in France (and the world) and published papers only of members (plus a few each year by members of the Académie at Montpellier which had a special relationship with the Paris Académie).

Of the 182 French (by regulation) articles published in the *Méms.,* Paris, 28% supported oxygen and 37% supported phlogiston, 15% were theoretical, 41% were quantitative, 22% used affinities, and 42% were written by men 40 or younger.

3. *Annales de Chimie, ou recueîl des mémoires concernant la chimie et les arts qui en dépendent,* started in 1789 but was suspended for a few years in 1793. Abbreviated herein *Ann. Chim.*

This was the major journal of the new chemistry, begun and edited by Lavoisier and other revolutionaries (compare Court, 1972). Since the *Mémoires* of the Académie published only members' papers and membership was severely restricted, and since the editor of the *Observations sur la Physique. . .* (Lamétherie) was opposed to the new paradigm, this provided the best outlet for young adherents of the new chemistry.

Of the 93 French articles published in *Ann. Chim.* (there was only 1 British), 91% supported oxygen and 0% phlogiston, 11% were theoretical, 44% were quantitative, 12% used affinities, and 75% were written by men 40 or younger. Since this journal was established as the organ of the revolutionaries, comparison with other journals without the time factor is almost meaningless. The very high

[10]Compare the discussion of the reasons for the publication of Crell's *Chemische Annalen* in Germany given by Hufbauer (1971, Chap. 4).

number of authors under 40 strongly supports the contention that this was a convenient outlet for young adherents of oxygen chemistry after 1789.

4. *Philosophical Transactions,* Royal Society, London, started in 1665. Abbreviated herein *Phil. Trans.*

This was the premier publication of the major scientific institution in Great Britain. During this period its reputation was not as high as it had been (Trengove, 1965), and many of the reviewing journals, especially in Germany, were quite critical of the poor quality of the papers it contained. However, at least as far as chemistry was concerned, some very important papers were published here, although they do not seem to be of as consistently high quality as those in the *Mémoires* of the Paris Académie.

Of the 65 British articles published in the *Phil. Trans.* (there were 2 French), 8% supported oxygen and 43% phlogiston, 21% were theoretical, 29% were quantitative, 15% used affinities, and 27% were written by authors 40 or younger. This is and should be typical of the British pattern of the distributions in Chapter 4, since this journal published 64% of all British articles.

5. *Mémoires de Mathématique et de Physique, Présentés a l'Académie Royale des Sciences, par Divers Savans, et Lus dans ses Assemblées,* also called *Savans Étrangers,* published from 1750 through 1786 in 11 volumes. Abbreviated herein *Savans Etr.*

This was a secondary publication of the Académie: it published articles by nonmembers and correspondents, lesser articles by members, and articles for special purposes, such as prize competitions. Although it was not restricted to French authors, there were no chemical articles by British authors published in it. Publication was irregular, averaging only one volume every four years.

Of the 35 French articles published in *Savans Etr.,* only 3% supported oxygen and 43% supported phlogiston, only 3% were theoretical, 34% were quantitative, 17% used affinities, and 65% were written by authors 40 or younger. The low percentage supporting oxygen is obviously due to the fact that it stopped publication just as the new theory was gaining favor. The low percentage theoretical is an indication of the second-class status of this journal, although the percentage quantitative was higher than that of the *Obs. Phys.* The high percentage of articles by young authors supports the assumption that this was indeed an outlet for beginning young chemists.

6. *Journal de Médecine, chirurgie, pharmacie, etc.,* published monthly in Paris from 1754 through 1794. Abbreviated herein *Jour. Méd.*

This journal, as its title indicates, was primarily devoted to medicine, and the chemical articles it contained were normally very concrete and applied in nature, and they concentrated on chemical knowledge useful for the medical practice. Still, there were occasional articles of wider import (in particular, Berthollet published some of his early articles there).

Of the 33 French articles published in *Jour. Méd.* (no British articles), none supported oxygen and only 12% supported phlogiston (fully 88% were "no information"), none were theoretical (91% were "descriptive" as befits such an applied journal), 9% were quantitative, 6% used affinities, and 38% were written by men 40 or younger.

7. *Mémoires* and *Nouveaux Mémoires,* Académie des Sciences, Arts, et Belles-lettres, Dijon, published from 1769 to 1772 and (*Nouveaux*) from 1782 to 1786. Abbreviated herein *Méms.,* Dijon.

This was the major publication, both in quantity and quality, in France outside of Paris. This is not surprising for Dijon was the major city of the Burgundy area, an area important in French culture since the middle ages. One reason for its early demise may be that the most important members, such as Morveau, moved on to Paris.

Of the 27 French articles published in *Méms.,* Dijon (again no British because of restriction to members), 0% supported oxygen and 36% phlogiston, only 4% were theoretical, 21% were quantitative, 18% used affinities, and only 8% were written by men 40 or younger. In spite of the fact that Guyton published many of his articles here, the percentages of theoretical, quantitative, and oxygen were very low. The last percentage resulted from the journal ceasing publication in 1786, while the percentage theoretical and quantitative may reflect some backwardness relative to the events going on in Paris. The very low percentage by younger authors is largely due to Guyton's dominance and late start and probably reflects the absence of chemical recruitment and career opportunities in Dijon (as further evidenced by the early demise of the *Mémoires*).

JOURNAL COMPARISONS

I will begin by making some overall comparisons among the journals before turning to a consideration of the time dimension. The only journals that had significant proportions of oxygen articles were *Ann. Chim.* with 90%, *Méms.,* Paris with 28%, and *Obs. Phys.* with 12% (French articles). The most theoretical of the French journals were *Obs. Phys.* and *Méms.,* Paris, both with 15%, followed by *Ann Chim.* with 11%. The least theoretical was *Jour. Méd.* with no theoretical articles while *Savans Etr.* and *Méms.,* Dijon had only a few. The most quantitative French journals were, as would be expected from the process of the revolution, *Ann. Chim.* with 44%, followed by *Méms.,* Paris with 41%. *Obs. Phys.* (French and British articles) and *Phil. Trans.* were less so, and *Jour. Méd.* was the least quantitative.

We look at the ages of authors in the journals as an indication of opportunity provided for new, young chemists. As expected from the nature of the Académie, less than half of the articles in *Méms.,* Paris were written by men under 40. Somewhat surprisingly, the same percentage of the French articles and only 9% of the British articles in *Obs. Phys.* were written by men of 40 or younger. Almost no young authors published in *Méms.,* Dijon or *Phil. Trans.* By contrast, 65% of the articles of the *Savans Etr.* were written by men under 40, as were 75% of the articles in the journal of the new chemistry, *Ann. Chim.* The fact that there was such a difference between the percentages of French and British authors under 40 publishing in *Obs. Phys.* shows that it was not so much the differing availability of publishing outlets in Britain and France (the *Phil. Trans.* was certainly easier to gain access to than *Méms.,* Paris) as the different social structures of the two scientific communities that led to disparities in the number of authors publishing and articles published. The journals merely manifested the situations in the two countries.

Reflecting the data in Chapters 3, 4, and 5, articles in *Phil. Trans.* and the British articles in *Obs. Phys.* exhibit similar characteristics. In relation to the French articles, they were more theoretical, slightly less quantitative, rarely supported the oxygen paradigm, and were only infrequently written by men under 40. Among the French journals, *Ann. Chim.,* as would be expected from its special history, published the most quantitative, pro-oxygen articles by the youngest authors of any major journal. It was also slightly lower in percentage theoretical than the other two major French journals, a result, according to our theory, of the fact that it had only begun publishing after the virtually certain success of the new paradigm in France.

In order to compare seriously the major journals in terms of the major variables (paradigm choice, theoretical level, and level of quantification) we should study their distribution over time. This is particularly important, both because we have hypothesized varying relationships between time and some of these variables, and because there were journals that were not published for identical times. For example, *Ann. Chim.* began publishing very late (1789) while *Savans Etr.* and *Méms.,* Dijon stopped early (1786).

Taken collectively, Tables B.3-B.5 reinforce the comparisons that have already been made among the journals. With the exception of periods 1778-80 and 1789-90, *Phil. Trans.* published a greater proportion of theoretical articles than any of the French journals for all periods, not including the British articles in *Obs. Phys.* (which were too variable to tell us much). Even when time is taken into account, we see that the *Ann. Chim.* produced articles more quantitative than the others, although it was joined by the *Méms.,* Paris after 1788 in supporting oxygen completely. The predilections of the editor of the *Obs. Phys.* are brought out clearly: this journal, in which the first antiphlogiston articles were published, soon lost its lead to the *Méms.,* Paris (and later, the *Ann. Chim.*) in percentage of articles supporting oxygen and percentage quantitative. The variation of these variables over time displays some initial support for our hypotheses concerning the relationships between time and oxygen, quantification, and theoretical level: oxygen and quantification rise while level of theory rises and then declines over time. The lesser journals with articles using the oxygen paradigm were *Mémoires,* Toulouse (2), *Médecine Eclairée* (9), *Journal des Mines* (7), and the journals of the Société d'Agriculture (1) and Société Royale de Médecine (2). Publication of theoretical or quantitative articles by any of these journals was too infrequent to be conclusive about anything except that they were minor journals.

TABLE B.3
Percentage of Articles in Each Period
which Supported Oxygen, by Journal

	1772 -77	1778 -80	1781 -84	1785 -86	1787 -88	1789 -90	1791 -95
			France				
Obs. Phys.	10 $(3)^a$	0 (0)	0 (0)	42 (8)	60 (6)	20 (3)	36 (4)
Méms., Paris	0 (0)	40 (8)	33 (9)	61 (11)	83 (15)	100 (4)	100 (4)
Ann. Chim.	$-^b$ —	— —	— —	— —	— —	100 (35)	100 (50)
			Britain				
Phil. Trans.	0 (0)	0 (0)	0 (0)	0 (0)	11 (1)	33 (3)	14 (1)

^{a}( ) = Number of articles. By "oxygen" we refer to our three-category scale; that is, these are articles which were originally classified antiphlogiston, new nomenclature, or oxygen. The reader should bear in mind that this scale does not include the "no information" articles. Therefore, the percentages cannot be compared with those given in the descriptions of the journals but only with others in the table.

^{b} — = Journal not published.

TABLE B.4
Percentage of Articles in Each Period
which Were Theoretical, by Journal

	1760 -65	1766 -71	1772 -77	1778 -80	1781 -84	1785 -86	1787 -88	1789 -90	1791 -95
				France					
Obs. Phys.	$-^a$ —	0 $(0)^b$	9 (5)	13 (3)	18 (4)	29 (9)	21 (3)	26 (5)	3 (1)
Méms., Paris	0 (0)	0 (0)	7 (1)	16 (5)	27 (11)	20 (5)	13 (4)	0 (0)	14 (1)
Ann. Chim.	— —	— —	— —	— —	— —	— —	— —	5 (2)	14 (8)
				Britain					
Obs. Phys	— —	0 (0)	0 (0)	0 (0)	0 (0)	80 (4)	0 (0)	50 (2)	0 (0)
Phil. Trans.	0 (0)	11 (1)	30 (3)	0 (0)	39 (5)	33 (1)	22 (2)	11 (1)	14 (1)

^{a} — = Journal not published.

^{b}( ) = Number of articles. The percentages refer to the highest category of the three-category scale. Means calculated from the 8-point scale exhibit similar patterns.

TABLE B.5
Percentage of Articles in Each Period
which Were Quantitative, by Journal

	1760 -65	1766 -71	1772 -77	1778 -80	1781 -84	1785 -86	1787 -88	1789 -90	1791 -95
					France				
Obs. Phys.	$-^a$	0	31	46	14	16	36	26	31
	−	$(0)^b$	(17)	(11)	(3)	(5)	(5)	(5)	(9)
Méms., Paris	25	25	47	59	42	52	39	0	0
	(2)	(4)	(7)	(19)	(17)	(13)	(12)	(0)	(0)
Ann. Chim.	−	−	−	−	−	−	−	41	45
	−	−	−	−	−	−	−	(15)	(25)
					Britain				
Obs. Phys.	−	0	0	0	0	80	20	25	0
	−	(0)	(0)	(0)	(0)	(4)	(1)	(1)	(0)
Phil. Trans.	0	33	0	67	54	33	11	33	29
	(0)	(3)	(0)	(2)	(7)	(1)	(1)	(3)	(2)

^{a} − = Journal not published.

^{b} ( ) = Number of articles.

Appendix C:
Supplementary Tables

TABLE C.1
Type of Theory over Time by Country
Basic Sample
(Percentages)

	1760 -71	1772 -77	1778 -80	1781 -84	1785 -86	1787 -88	1789 -90	1791 -95	Total
				France					
Phlogiston	40	50	40	39	33	12	9	4	29
Old nomenclature	–	–	–	–	–	0	9	3	2
Neutral	1	0	0	0	2	0	1	1	1
Antiphlogiston	0	3	9	2	7	4	4	0	3
New nomenclature	–	–	–	–	–	12	36	37	11
Oxygen	0	0	0	7	16	30	25	29	13
No information	59	47	52	53	41	42	16	26	42
Total	100	100	101^a	101^a	99^a	100	100	99^a	101^a
Number of cases	(75)	(110)	(91)	(104)	(85)	(57)	(76)	(126)	(724)
				Britain					
Phlogiston	10	33	14	71	78	42	50	10	37
Old nomenclature	–	–	–	–	–	21	8	33	10
Neutral	0	0	0	0	0	0	4	0	1
Antiphlogiston	0	0	0	0	0	0	0	5	1
New nomenclature	–	–	–	–	–	5	12	5	4
Oxygen	0	0	0	0	0	0	0	10	1
No information	90	67	86	29	22	32	27	38	47
Total	100	100	100	100	100	100	101^a	101^a	101^a
Number of cases	(20)	(18)	(7)	(14)	(9)	(19)	(26)	(21)	(134)

^{a}Differs from 100 due to rounding.

TABLE C.2
Level of Theory over Time by Country
Basic Sample
(Percentage Distribution)

	1760 -71	1772 -77	1778 -80	1781 -84	1785 -86	1787 -88	1789 -90	1791 -95	Total
				France					
Descriptive	65	56	54	39	34	44	58	58	51
Mixed	32	34	35	46	49	42	33	34	38
Theoretical	3	10	11	14	16	14	9	8	11
Total	100	100	100	99[a]	99[a]	100	100	100	100
Number of cases	(75)	(110)	(91)	(104)	(85)	(57)	(76)	(126)	(724)
				Britain					
Descriptive	55	44	57	21	11	53	35	71	46
Mixed	40	39	43	43	33	37	54	24	40
Theoretical	5	17	0	36	56	11	12	5	15
Total	100	100	100	100	100	101[a]	101[a]	100	101[a]
Number of cases	(20)	(18)	(7)	(14)	(9)	(19)	(26)	(21)	(134)

[a]Differs from 100 due to rounding.

TABLE C.3
Level of Quantification over Time by Country
Basic Sample

	1760 -71	1772 -77	1778 -80	1781 -84	1785 -86	1787 -88	1789 -90	1791 - 95	Total
				France					
Quantitative (%)	16	25	40	27	31	35	33	39	31
Number of cases	(75)	(110)	(91)	(104)	(85)	(57)	(76)	(126)	(724)
				Britain					
Quantitative (%)	25	0	43	50	56	21	31	43	31
Number of cases	(20)	(18)	(7)	(14)	(9)	(19)	(26)	(21)	(134)

TABLE C.4
Distribution of Oxygen Articles Over Time
by Level of Theory by Country
Basic Sample

	1760 -71	1772 -77	1778 -80	1781 -84	1785 -86	1787 -88	1789 -90	1791 -95	Total
				France					
Nontheoretical (%)	0	0	4	4	32	79	79	88	
Number of cases	(25)	(37)	(27)	(24)	(25)[a]	(19)	(52)	(64)	(273)
Theoretical (%)	0	14	41	32	48	79	67	93	
Number of cases	(6)*	(21)	(17)	(25)	(25)[a]	(14)	(12)[a]	(29)[a]	(149)
				Britain					
Nontheoretical (%)	0	0	0	0	0	10	30	13	14
Number of cases	(2)	(2)	(1)	(1)	(1)	(10)	(10)[a]	(8)[b]	(35)
Theoretical (%)	—	0	—	0	0	0	0	80	11
Number of cases	—	(4)	—	(9)	(6)	(3)	(9)	(5)	(36)

[a] Including 1 neutral.

[b] Including 5 neutral.

TABLE C.5
Percentage Distribution of Oxygen Articles over Time by
Level of Quantification by Country
Basic Sample

	1760 -71	1772 -77	1778 -80	1781 -84	1785 -86	1787 -88	1789 -90	1791 -95	Total
				France					
Qualitative (%)	0	0	4	6	27	69	72	86	36
Number of cases	(27)[a]	(39)	(26)	(34)	(30)[b]	(16)	(43)[a]	(51)[a]	(266)
Quantitative (%)	0	16	39	47	60	88	86	93	65
Number of cases	(4)	(19)	(18)	(15)	(20)	(17)	(21)	(42)	(156)
				Britain					
Qualitative (%)	0	0	0	0	0	10	25	33	15
Number of cases	(1)	(6)	(1)	(5)	(2)	(10)	(12)[a]	(9)[c]	(46)
Quantitative (%)	0	—	—	0	0	0	0	50	8
Number of cases	(1)	(0)	(0)	(5)	(5)	(3)	(7)	(4)[b]	(25)

[a] Including 1 neutral article.

[b] Including 2 neutral articles.

[c] Including 3 neutral articles.

TABLE C.6

Percentage Distribution of Oxygen Articles over Time by Level of Theory
France: Analytic Sample, Lavoisier Omitted

	1760 -71	1772 -77	1778 -80	1781 -84	1785 -86	1787 -88	1789 -90	1791 -95
Nontheoretical (%)	0	0	0	0	17	73	79	88
Number of cases	(15)	(30)	(17)	(22)	(12)[a]	(15)	(43)	(51)
Theoretical (%)	0	17	0	11	28	73	67	96
Number of cases	(5)[a]	(12)	(6)	(18)	(18)[a]	(11)	(9)	(25)

[a]Includes 1 neutral article.

TABLE C.7

Percentage Distribution of Oxygen Articles over Time by Level of Quantification
France: Analytic Sample, Lavoisier Omitted

	1760 -71	1772 -77	1778 -80	1781 -84	1785 -86	1787 -88	1789 -90	1791 -95
Qualitative (%)	0	0	0	0	25	64	76	90
Number of cases	(17)[a]	(28)	(16)	(30)	(20)[b]	(14)	(38)	(42)
Quantitative (%)	0	14	0	20	20	83	79	91
Number of cases	(3)	(14)	(7)	(10)	(10)	(12)	(14)	(34)

[a]Includes 1 neutral article.

[b]Includes 2 neutral articles.

TABLE C.8
Percentage Distribution of Oxygen Articles over Time by Age, Controlling For Country,
Basic Sample

	1760 -71	1772 -77	1778 -80	1781 -84	1785 -86	1787 -88	1789 -90	1791 -95	Total
					France				
Forty or younger (%)	0	5	44	14	35	100	86	100	
Number of cases	(10)[a]	(21)	(18)	(21)	(20)[a]	(16)	(37)	(59)	(202)
Forty-one or older (%)	0	7	0	23	57	53	62	64	
Number of cases	(14)	(29)	(14)	(26)	(23)[a]	(15)	(21)	(22)	(164)
					Britain				
Forty or younger (%)	–	0	–	–	–	0	0	25	10
Number of cases	(0)	(2)	(0)	(0)	(0)	(2)	(2)	(4)[b]	(10)
Forty-one or older (%)	0	0	0	0	0	0	13	43	9
Number of cases	(2)	(4)	(1)	(10)	(7)	(9)	(15)[a]	(7)[a]	(55)

[a]Including 1 neutral.

[b]Including 3 neutral.

TABLE C.9
Percentage Distribution of Oxygen Articles over Time by Age
France: Analytic Sample, Lavoisier Included

	1760 -71	1772 -77	1778 -80	1781 -84	1785 -86	1787 -88	1789 -90	1791 -95
Forty or younger (%)	0	6	47	14	35	100	86	100
Number of cases	(8)[a]	(18)	(17)	(21)	(20)[a]	(16)	(19)	(59)
Forty-one or older (%)	0	7	0	23	57	50	61	63
Number of cases	(13)	(28)	(14)	(26)	(23)[a]	(14)	(18)	(19)

[a]Includes 1 neutral article.

TABLE C.10
Means and Standard Deviations for Path Analyses

Variables	1760-77	1778-84	1785-88	1789-95	Overall
			France		
Time	12.4	21.5	26.4	30.9	24.3
	$(4.3)^a$	(2.3)	(1.2)	(1.6)	(7.2)
Level of theory	3.7	4.4	4.8	3.3	3.9
	(2.1)	(2.3)	(2.2)	(2.4)	(2.3)
Level of	1.3	1.4	1.5	1.4	1.4
quantification	(0.5)	(0.5)	(0.5)	(0.5)	(0.5)
Age	43.9	41.7	39.6	36.9	40.0
	(10.4)	(9.7)	(6.9)	(8.6)	(9.4)
Phlogiston/	1.1	1.4	2.2	2.7	2.0
Oxygen	(0.4)	(0.8)	(1.0)	(0.7)	(1.0)
Number of cases	67	78	73	133	351
			Britain		
Time	13.9	22.7	26.7	30.2	25.8
	(1.9)	(1.6)	(1.4)	(1.3)	(5.4)
Level of theory	4.7	5.5	4.6	4.3	4.7
	(2.6)	(2.4)	(2.6)	(2.3)	(2.4)
Level of	1.1	1.5	1.4	1.3	1.3
quantification	(0.4)	(0.5)	(0.5)	(0.5)	(0.5)
Age	41.7	49.3	53.3	50.3	49.9
	(1.7)	(3.5)	(4.0)	(10.7)	(8.0)
Phlogiston/	1.0	1.0	1.0	1.6	1.2
Oxygen	(0.0)	(0.0)	(0.0)	(0.8)	(0.6)
Number of cases	7	11	17	23	58

[a]( ) = Standard deviation. The coding of the variables is as follows: Time (0-35, i.e., 1760-95 − add 60 to any value to get the year); Level of theory (1-8); quantitativeness (1-2); age (18-90, actual age of the author); and phlogiston/oxygen (1 − phlogiston, 2 − neutral, 3 − oxygen).

TABLE C.11
Correlation Matrix for Basic Path Model
(France and Britain Combined)

	Level of theory	Level of quantification	Phlogiston/ oxygen	Age	Country
Time	−.04	.11	.55	−.19	−.07
Theory		.09	.01	.00	−.11
Quantification			.19	−.01	.03
Oxygen				−.40	.28
Age					−.36
Number of cases = 409					

TABLE C.12
Correlation Matrices for Path Models of Chapter Five

	Level of theory	Level of quantification	Phlogiston/ oxygen	Age
		France		
		1760-1777		
Time	.16	.29	.08	−.02
Theory		.27	.41	−.13
Quantification			.29	.06
Oxygen				−.04
Number of cases = 67				
		1778-1784		
Time	.23	.02	.04	−.18
Theory		.18	.47	−.33
Quantification			.49	−.02
Oxygen				−.20
Number of cases = 78				
		1785-1788		
Time	−.13	.01	.30	.05
Theory		−.02	.07	.02
Quantification			.24	.02
Oxygen				−.12
Number of cases = 73				
		1789-1790		
Time	−.10	.27	.10	−.01
Theory		−.04	.03	−.02
Quantification			.06	.00
Oxygen				−.37
Number of cases = 133				
		Overall		
Time	−.04	.12	.62	−.28
Theory		.08	.07	−.07
Quantification			.22	−.01
Oxygen				−.34
Number of cases = 351				
		Britain		
		1760-1777		
Time	.81	−.68	−	.19
Theory		−.29	−	.50
Quantification			−	.59
Oxygen			−	−
Number of cases = 7				

(continued)

TABLE C.12 *(continued)*

	Level of theory	Level of quantification	Phlogiston/ oxygen	Age
		1778-1784		
Time	.69	.04	−	.59
Theory		.10	−	.78
Quantification			−	−.35
Oxygen				−
Number of cases = 11				
		1785-1788		
Time	−.52	−.52	−	.12
Theory		.46	−	.05
Quantification			−	.03
Oxygen				−
Number of cases = 17				
		1789-1795		
Time	−.32	−.02	.40	−.30
Theory		.01	−.27	.18
Quantification			−.11	.05
Oxygen				−.45
Number of cases = 23				
		Overall		
Time	−.10	.01	.37	.31
Theory		.16	−.20	.15
Quantification			−.09	.08
Oxygen				−.32
Number of cases = 58				

TABLE C.13
Standardized and Unstandardized Regression Coefficients for Models
of Revolution, Cited Sample[a]

Independent variables	Dependent variables							
	1760-1777		1778-1784		1785-1788		1789-1795	
	France							
	Level of theory							
Time	.05	$(.12)^b$	.21	$(.19)^c$	−.14	(−.07)	−.18	(−.13)
R-squared	.01		.03		.00		.02	
	Level of quantification							
Time	.01	(.15)	−.03	(−.12)	.02	(.06)	.06	$(.21)^c$
Theory	.06	$(.25)^c$	.03	(.13)	.04	$(.22)^c$	.01	(.05)
R-squared	.09		.03		.05		.04	
	Phlogiston/oxygen							
Time	−.01	(−.13)	.00	(.00)	.16	$(.25)^c$	.02	(.05)
Theory	−.07	$(−.25)^c$	.03	(.09)	.02	(.07)	.03	(.11)
Quantification	.13	(.11)	.40	$(.29)^c$	.34	$(.22)^c$	.10	(.07)
Age	.00	(−.03)	.00	(−.02)	−.01	(−.11)	−.03	$(−.43)^c$
R-squared	.08		.10		.13		.22	
$R_{\text{age-time}}$	.19		−.09		.12		.10	
Number of cases	131		150		123		171	
	Britain							
	Level of Theory							
Time	.16	(.28)	.64	$(.53)^c$	−.99	$(−.52)^c$	−.39	(−.29)
R-squared	.08		.28		.27		.09	
	Level of quantification							
Time	−.01	(−.18)	−.02	(−.08)	−.16	(−.43)	.05	(.17)
Theory	−.02	(−.14)	.04	(.20)	.03	(.14)	.00	(.00)
R-squared	.07		.03		.27		.03	
	Phlogiston/oxygen							
Time	−.05	$(−.42)^c$	−.02	(−.07)	.02	(.07)	.11	(.26)
Theory	−.04	(−.23)	−.13	$(−.63)^c$	−.04	(−.24)	−.07	(−.21)
Quantification	−.24	−.17)	.12	(.12)	−.05	(−.06)	−.04	(−.02)
Age	.00	(.08)	.01	(.13)	.00	(.00)	−.02	(−.29)
R-squared	.27		.41		.10		.26	
$R_{\text{age-time}}$	.04		.10		.09		−.08	
Number of cases	27		19		22		31	

[a]"No information" assumed neutral.
[b]( ) = standard coefficient.
[c]$p < .05$.

TABLE C.14
Standardized and Unstandardized Regression Coefficients for Models
of Revolution, Cited Sample[a]

Independent variables	Dependent variables			
	1760-1777	1778-1784	1785-1788	1789-1795
	France			
	Level of theory			
Time	.05 (.12)[b]	.24 (.22)[c]	−.03 (−.01)	−.17 (−.12)
R-squared	.01	.05	.00	.01
	Level of quantification			
Time	.01 (.15)	−.03 (−.11)	.03 (.07)	.07 (.22)[c]
Theory	.06 (.25)[c]	.03 (.14)	.04 (.19)[c]	.00 (.04)
R-squared	.09	. .03	.04	.05
	Phlogiston/oxygen			
Time	.00 (.04)	−.06 (−.14)	.22 (.27)[c]	.06 (.10)
Theory	.05 (.27)[c]	.10 (.28)[c]	.05 (.12)	.06 (.17)
Quantification	.16 (.20)[c]	.74 (.44)[c]	.66 (.32)[c]	.13 (.07)
Age	.00 (.01)	.00 (−.01)	−.02 (−.17)	−.05 (−.50)[c]
R-squared	.15	.32	.24	.32
$R_{\text{age-time}}$	.19	−.12	.07	.10
Number of cases	131	140	115	166
	Britain			
	Level of theory			
Time	.16 (.28)	.72 (.60)[c]	−1.0 (−.52)[c]	−.51 (−.39)[c]
R-squared	.08	.36	.27	.15
	Level of quantification			
Time	−.01 (−.18)	−.08 (−.34)	−.16 (−.43)	.02 (.07)
Theory	−.02 (−.14)	.09 (.45)	.03 (.14)	−.01 (−.06)
R-squared	.07	.13	.27	.01
	Phlogiston/oxygen			
Time	— —	— —	— —	.12 (.23)
Theory	— —	— —	— —	−.07 (−.18)
Quantification	— —	— —	— —	−.15 (−.08)
Age	— —	— —	— —	−.06 (−.55)[c]
R-squared	—	—	—	.46
$R_{\text{age-time}}$	.04	.16	.09	−.20
Number of cases	27	17	22	28

[a]Best guess for "no information."

[b]( ) = standardized coefficient.

[c]$p < .05.$

TABLE C.15
Standardized and Unstandardized Regression Coefficients for Models
of Revolution, Basic Sample[a]

Independent variables	Dependent variables							
	1760-1777		1778-1784		1785-1788		1789-1795	
France								
Level of theory								
Time	.09	$(.17)^b$	.23	$(.23)^c$	−.27	(−.15)	−.13	(−.09)
R-squared	.03		.06		.02		.01	
Level of quantification								
Time	.03	$(.26)^c$	.00	(−.02)	.00	(.02)	.08	$(.25)^c$
Theory	.03	(.14)	.04	(.20)	.00	(−.03)	.00	(−.02)
R-squared	.10		.04		.00		.06	
Phlogiston/oxygen								
Time	.00	(−.04)	−.02	(−.07)	.27	$(.33)^c$	.04	(.08)
Theory	.06	$(.31)^c$	.14	$(.38)^c$	.05	(.12)	.01	(.05)
Quantification	.23	$(.25)^c$	.72	$(.42)^c$	.49	$(.25)^c$	.13	(.09)
Age	.00	(.00)	.00	(−.08)	−.02	(−.14)	−.03	$(−.38)^c$
R-squared	.18		.40		.19		.17	
$R_{age-time}$	−.04		−.17		.05		−.03	
Number of cases	74		79		74		139	
Britain								
Level of theory								
Time	.93	$(.80)^c$	1.02	$(.69)^c$	−.92	$(−.50)^c$	−.67	$(−.42)^c$
R-squared	.64		.48		.25		.17	
Level of quantification								
Time	−.11	(−.66)	−.02	(−.05)	−.15	(−.43)	.01	(.04)
Theory	.04	(.32)	.03	(.14)	.04	(.23)	.03	(.12)
R-squared	.20		.01		.33		.01	
Phlogiston/oxygen								
Time	—	—	—	—	—	—	.26	$(.43)^c$
Theory	—	—	—	—	—	—	−.03	(−.07)
Quantification	—	—	—	—	—	—	−.05	(−.03)
Age	—	—	—	—	—	—	−.01	(−.20)
R-squared	—		—		—		.28	
$R_{age-time}$	−.22		.59		−.09		−.09	
Number of cases	9		11		18		28	

[a] "No information" omitted.

[b] ( ) = standardized coefficient.

[c] $p < .05$.

TABLE C.16
Standardized and Unstandardized Regression Coefficients for Models
of Revolution, Basic Sample[a]

Independent variables	Dependent variables							
	1760-1777		1778-1784		1785-1788		1789-1795	
France								
Level of theory								
Time	.05	$(.13)^b$	.21	(.19)	−.14	(−.07)	−.14	(−.10)
R-squared	.02		.04		.01		.01	
Level of quantification								
Time	.01	(.15)	−.03	(−.12)	.03	(.07)	.06	$(.20)^c$
Theory	.04	$(.19)^c$	.03	(.13)	.04	$(.21)^c$	.01	(.05)
R-squared	.07		.03		.05		.04	
Phlogiston/oxygen								
Time	−.01	(−.12)	.00	(.00)	.16	$(.26)^c$	.02	(.05)
Theory	−.08	$(−.29)^c$	.03	(.10)	.02	(.07)	.03	(.11)
Quantification	.15	(.12)	.41	$(.30)^c$	.35	$(.22)^c$	.16	(.11)
Age	.00	(−.03)	.00	(.00)	−.01	(−.11)	−.03	$(−.43)^c$
R-squared	.10		.10		.14		.24	
$R_{age-time}$	.17		−.09		.12		.08	
Number of cases	142		151		124		179	
Britain								
Level of theory								
Time	.19	$(.35)^c$	.64	$(.55)^c$	−.98	$(−.48)^c$	−.39	(−.30)
R-squared	.12		.30		.23		.09	
Level of quantification								
Time	−.02	(−.27)	−.01	(−.05)	−.15	(−.40)	.06	(.22)
Theory	−.02	(−.11)	.02	(.09)	.03	(.17)	.03	(.13)
R-squared	.10		.01		.25		.05	
Phlogiston/oxygen								
Time	−.04	$(−.38)^c$	−.05	(−.19)	−.05	(−.13)	.13	(.31)
Theory	−.04	(−.23)	−.11	$(−.52)^c$	−.06	(−.34)	−.07	(−.22)
Quantification	−.20	(−.15)	.07	(.07)	−.19	(−.19)	−.04	(−.03)
Age	.00	(.07)	.00	(.05)	.00	(−.09)	−.01	(−.17)
R-squared	.23		.39		.16		.21	
$R_{age-time}$	.14		.20		.01		.03	
Number of cases	34		21		26		39	

[a]"No information" assumed neutral.

[b]( ) = standardized coefficient.

[c]$p < .05$.

TABLE C.17
Standardized and Unstandardized Regression Coefficients for Models
of Revolution, Basic Sample[a]

Independent variables	Dependent variables							
	1760-1777		1778-1784		1785-1788		1789-1795	
	France							
	Level of theory							
Time	.05	$(.13)^b$	.24	$(.22)^c$	−.04	(−.02)	−.13	(−.10)
R-squared	.02		.05		.00		.01	
	Level of quantification							
Time	.01	(.15)	−.03	(−.11)	.03	(.08)	.06	$(.21)^c$
Theory	.04	$(.19)^c$	.03	(.14)	.04	$(.19)^c$	.00	(.03)
R-squared	.07		.03		.04		.04	
	Phlogiston/oxygen							
Time	.00	(.04)	−.06	(−.14)	.22	$(.27)^c$	.06	(.10)
Theory	.04	$(.24)^c$	.10	$(.28)^c$	.05	(.12)	.07	(.17)
Quantification	.18	$(.23)^c$	.73	$(.44)^c$	.67	$(.32)^c$	.20	(.11)
Age	.00	(.02)	.00	(−.01)	−.02	$(−.17)^c$	−.05	$(−.50)^c$
R-squared	.14		.32		.25		.33	
$R_{age-time}$	.17		−.11		.07		.08	
Number of cases	142		141		116		174	
	Britain							
	Level of theory							
Time	.19	$(.35)^c$	.72	$(.62)^c$	−.94	$(−.50)^c$	−.44	(−.34)
R-squared	.12		.38		.25		.11	
	Level of quantification							
Time	−.02	(−.28)	−.07	(−.29)	−.17	$(−.46)^c$	.05	(.18)
Theory	−.02	(−.11)	.06	(.30)	.02	(.12)	.00	(.03)
R-squared	.10		.07		.28		.03	
	Phlogiston/oxygen							
Time	−	−	−	−	−	−	.23	$(.44)^c$
Theory	−	−	−	−	−	−	−.02	(−.04)
Quantification	−	−	−	−	−	−	−.08	(−.05)
Age	−	−	−	−	−	−	−.03	$(−.33)^c$
R-squared	−		−		−		.32	
$R_{age-time}$	.14		.25		−.04		−.05	
Number of cases	34		19		23		33	

[a]Best guess for "no information."

[b]( ) = standardized coefficient

[c]$p < .05$.

References

Académie des Sciences, Paris. *Index Biographique des Membres et Correspondents de l'Académie des Sciences de 1666 à 1939.* Paris: Gauthier-Villars, 1939.

Althauser, R. P. Multicollinearity and non-additive regression models. In H. M. Blalock (Ed.), *Causal models in the social sciences.* Chicago: Aldine, 1971. Pp. 453–472.

Bayer, A. E., & Folger, J. Some correlates of a citation measure of productivity in science. *Sociology of Education,* 1966, *39,* 381–390.

Bedel, C. L'enseignement des sciences pharmaceutiques. In R. Taton (Ed.), *Enseignement et diffusion des sciences en France au XVIIIe siècle.* Paris: Hermann, 1964. Pp. 237–258. (a)

Bedel, C. Les cabinets de chimie. In R. Taton (Ed.), *Enseignement et diffusion des sciences en France au XVIIIe siècle.* Paris: Hermann, 1964. Pp. 647–652. (b)

Ben-David, J., & Sullivan, T. Sociology of science. In A. Inkeles, J. Coleman, & N. Smelser (Eds.), *Annual review of sociology.* Palo Alto: Annual Reviews, Inc., 1975. Pp. 203–222.

Birembaut, A. L'enseignement de la mineralogie et des techniques minières. In R. Taton (Ed.), *Enseignement et diffusion des sciences en France au XVIII e siècle.* Paris: Hermann, 1964. Pp. 365–418.

Blalock, H. M. *Causal inferences in nonexperimental research.* Chapel Hill, North Carolina: University of North Carolina Press, 1964.

Blalock, H. M. Causal inferences, closed populations, and measures of association. *Causal models in the social sciences.* Chicago: Aldine, 1971. Pp. 139–151. (Reprinted from *American Political Science Review,* 1967, *61,* 130–136.) (b)

Blalock, H. M. *Social statistics* (2nd ed). New York: McGraw-Hill, 1972.

Blume, S. S., & Sinclair, R. Chemists in British universities: A study in the reward system of science. *American Sociological Review,* 1973, *38,* 126–138.

Boas, M. Structure of matter and chemical theory in the seventeenth and eighteenth centuries. In M. Clagett (Ed.), *Critical problems in the history of science.* Madison, Wisconsin: University of Wisconsin Press, 1959. Pp. 499–514.

Boerhaave, H. *A new method of chemistry; including the history, theory, and practice of the art.* Translated from *Elementa chemiae* (2 vols., Leyder, 1732) by Peter Shaw. London: Longman, 1741.

Bohrnstedt, G., & Carter, T. M. Robustness in regression analysis. In H. Costner (Ed.), *Sociological methodology: 1971.* San Francisco: Jossey–Bass, 1971. Pp. 118–146.

Bolton, H. C. *A select bibilography of chemistry: 1492-1892.* Washington, D.C.: The Smithsonian Institution, 1893.

Boyle, R. P. Path analysis and ordinal data. *American Journal of Sociology,* 1970, *75,* 461–480.

Cain, G. & Watts, H. Problems in making policy inferences from the Coleman report. *American Sociological Review,* 1970, *35,* 228–241.

Cartwright, F. F. The association of Thomas Beddoes, M.D. with James Watt, F. R.S. *Notes and Records of the Royal Philosophical Society, London,* 1967, *22,* 131–141.

Chaldecott, J. A. Scientific activities in Paris in 1791. *Annals of Science,* 1968, *24,* 21–52.

Clow, A., & Clow, N. L. *The chemical revolution: A contribution to social technology.* London: The Batchworth Press, 1952.

Cochrane, R. C. Francis Bacon and the rise of the mechanical arts in eighteenth-century England. *Annals of Science,* 1957, *12,* 37–156.

Cole, S. Professional standing and the reception of scientific discoveries. *American Journal of Sociology,* 1970, *76,* 286–306.

Cole, S., & Cole, J. Scientific output and recognition: A study in the operation of the reward system in science. *American Sociological Review,* 1967, *32,* 377–390.

Coleby, L. J. M. John Mickleburgh, Professor of Chemistry in the University of Cambridge, 1718-56. *Annals of Science,* 1952, *8,* 165–174.

Coleby, L. J. M. Richard Watson, Professor of Chemistry in the University of Cambridge, 1764-71. *Annals of Science,* 1953, *9,* 101–123.

Coleby, L. J. M. Isaac Milner and the Jacksonian chair of natural philosophy. *Annals of Science,* 1954, *10,* 234–257.

Collins, R. Competition and social control in science: An essay in theory construction. *Sociology of Education,* 1968, *41,* 123–140.

Costabel, P. Institutions et structures. In R. Taton (Ed.), *Enseignement et diffusion des sciences en France au XVIIe siècle.* Paris: Hermann, 1964. Pp. 15–26. (a)

Costabel, P. L'Oratoire de France et ses collèges. In R. Taton (Ed.), *Enseignement et diffusion des sciences en France au XVIIIe siècle.* Paris: Hermann, 1964. Pp. 67–100. (b)

Court, S. The *Annales de chimie,* 1789-1815. *Ambix,* 1972, *19,* 113–128.

Crane, D. Scientists at major and minor universities: a study of productivity and recognition. *American Sociological Review,* 1965, *30,* 699–714.

Crane, D. Social structure in a group of scientists: A test of the 'invisible college' hypothesis. *American Sociological Review,* 1969, *34,* 335–351.

Crane, D. *Invisible colleges.* Chicago: University of Chicago Press, 1972.

Crosland, M. P. *Historical studies in the language of chemistry.* London: Heinemann, 1962.

Crosland, M. P. The development of chemistry in the eighteenth century. *Studies on Voltaire and the eighteenth century,* 1963, *24,* 369–441.

Crosland, M. P. *The Society of Arcueil.* Cambridge, Massachusetts: Harvard University Press, 1967.

Crosland, M. P. (Ed.), *Science in France in the revolutionary era.* Cambridge, Massachusetts: MIT Press, 1969.

Crosland, M. P. Lavoisier's Theory of Acidity. *Isis,* 1973, *64,* 306–326.

de Dainville, F. L'enseignement scientifique dans les collèges des Jesuites. In R. Taton (Ed.), *Enseignement et diffusion des sciences en France au XVIIIe siècle.* Paris: Hermann, 1964.

Daumas, M., & Duveen, D. Lavoisier's relatively unknown large-scale decomposition and synthesis of water, February 27 and 28, 1785. *Chymia,* 1959, *5,* 113–129.

Duncan, A. M. Some theoretical aspects of eighteenth-century tables of affinity – I. *Annals of Science,* 1964, *18,* 177–232.

Duncan, A. M. William Keir's *De Attractione Chemica* (1778) and the concepts of chemical saturation, attraction and repulsion. *Annals of Science,* 1967, *23,* 149–173.

Duncan, O. D. Path analysis: Sociological examples. *American Journal of Sociology,* 1966, *72,* 1–16.

Duveen, D., & Klickstein, H. A letter from Berthollet to Blagden relating to the experiments for a large-scale synthesis of water carried out by Lavoisier and Meusnier in 1785. *Annals of Science,* 1954, *10,* 58–62. (a)

Duveen, D., & Klickstein, H. A bibliographical study of the introduction of Lavoisier's *Traité Élémentaire de chimie* into Great Britain and America. *Annals of Science,* 1954, *10,* 321–338. (b)

Duveen, D., & Klickstein, H. A letter from Guyton de Morveau to Macquart relating to Lavoisier's attack against the *phlogiston* theory (1778); with an account of de Morveau's conversion to Lavoisier's doctrines in 1787. *Osiris,* 1956, *12,* 342–367.

Fichman, M. French Stahlism and eighteennth-century chemistry. Unpublished Ph.D. dissertation, Harvard University, 1969.

Fisher, C. S. The death of a mathematical theory. *Archive for the History of the Exact Sciences,* 1966, *3,* 137–159.

Fisher, C. S. The last invariant theorists. *European Journal of Sociology,* 1967, *8,* 216–244.

Fisher, C. S. Some social characteristics of mathematicians and their work. *American Journal of Sociology,* 1973, *78,* 1094–1118.

Fric, R. Contribution à l'étude de l'évolution des idées de Lavoisier sur la nature de l'air et sur la calcination des métaux. *Archives Internationales d'Histoire des Sciences,* 1959, *12,* 137–168.

Garrison, F. H. The medical and scientific periodicals of the 17th and 18th centuries. *Bulletin of the Institute of the History of Medicine,* 1934, *2,* 285–343.

Garvey, W. D., Lin, N., & Nelson, C. E. Communication in the physical and social sciences. *Science,* Dec. 11, 1970, 1166–1173.

Gaston, J. C. The reward system in British science. *American Sociological Review,* 1970, *35,* 718–732.

Gaston, J. C. Secretiveness and competition for priority of discovery in physics. *Minerva,* 1971, *9,* 472–492.

Gibbs, F. W. William Lewis, M. B., F. R. S. (1708-1781). *Annals of Science,* 1952, *8,* 122–151. (a)

Gibbs, F. W. Prelude to chemistry in industry. *Annals of Science,* 1952, *8,* 271–281. (b)

Gibbs, F. W., & Smeaton, W. A. Thomas Beddoes at Oxford. *Ambix,* 1961, *9,* 47–49.

Gillispie, C. C. The natural history of industry. *Isis,* 1957, *48,* 398–407.

Gillispie, C. C. The *Encyclopédie* and the Jacobin philosophy of science: a study in ideas and consequences. In M. Clagett (Ed.), *Critical problems in the history of science.* Madison, Wisconsin: University of Wisconsin Press, 1959. Pp. 255–290.

Gough, J. B. Lavoisier's early career in science: an examination of some new evidence. *British Journal for the History of Science,* 1968, *4,* 52–57.

Gough, J. B. Nouvelle contribution à l'étude de l'évolution des idées de Lavoisier sur la nature de l'air et sur la calcination des métaux. *Archives Internationales d'Histoire des Sciences,* 1969, *22,* 267–273.

Guerlac, H. The continental reputation of Stephen Hales. *Archives Internationales d'Histoire des Sciences,* 1951, *4,* 393–404.

Guerlac, H. A note on Lavoisier's scientific education. *Isis,* 1956, *47,* 211–216.

Guerlac, H. The origin of Lavoisier's work on combustion. *Archives Internationales d'Histoire des Sciences,* 1959, *12,* 113–115. (a)

Guerlac, H. Some French antecedents of the chemical revolution. *Chymia,* 1959, *5,* 73–112. (b)

Guerlac, H. Commentary on the papers of Charles Coulston Gillispie and L. Pearce Williams. In M. Clagett (Ed.), *Critical problems in the history of science.* Madison Wisconsin: University of Wisconsin Press, 1959. Pp. 317–320. (c)

Guerlac, H. Quantification in chemistry. *Isis*, 1961, *52*, 194–214. (a)

Guerlac, H. A curious Lavoisier episode. *Chymia*, 1961, *7*, 103–108. (b)

Guerlac, H. Lavoisier's draft memoir of July 1772. *Isis*, 1969, *60*, 380–382.

Gustin, B. H. Charisma, recognition, and the motivation of scientists. *American Journal of Sociology*, 1973, *78*, 1119–1134.

Hagstrom, W. O. *The scientific community.* New York: Basic Books, 1965.

Hahn, R. *The anatomy of a scientific institution: the Paris Academy of Sciences, 1666–1803.* Berkeley, California: The University of California Press, 1971.

Hales, S. *Vegetable staticks: Or, an account of stome statical experiments.* . .London: Innys & Woodward, 1727.

Hans, N. *New trends in education in the eighteenth century.* London: Routledge & Kegan Paul, 1951.

Hempel, C. *Aspects of scientific explanation.* New York: Free Press, 1964.

Heise, D. R. Problems in path analysis and causal inference. In E. F. Borgatta & G. W. Bohrnstedt (Eds.), *Sociological methodology: 1969.* San Francisco: Jossey-Bass, 1969. Pp. 38–73.

Hill, H. B. Commentary on the papers of Charles Coulston Gillispie and L. Pearce Williams. In M. Clagett (Ed.), *Critical problems in the history of science,* Madison, Wisconsin: University of Wisconsin Press, 1959. Pp. 309–316.

Huard, P. L'Enseignement medico-chirurgical. In R. Taton (Ed.), *Enseignement et diffusion des sciences en France au XVIIIe siècle.* Paris: Hermann, 1964. Pp. 171–236.

Hufbauer, K. G. The formation of the German chemical community. Unpublished doctoral dissertation, University of California, Berkeley, 1971.

Johnston, J. *Econometric methods.* New York: McGraw-Hill, 1972.

Kohler, R. E., Jr. The origin of Lavoisier's first experiments on combustion. *Isis*, 1972, *63*, 349–355.

Kuhn, T. S. Robert Boyle and structural chemistry in the seventeenth century. *Isis*, 1952, *43*, 12–36.

Kuhn, T. S. The function of measurement in modern physical science. *Isis*, 1961, *52*, 169–193.

Kuhn, T. S. *The structure of scientific revolutions.* Chicago: University of Chicago Press, 1962.

Kuhn, T. S. *The structure of scientific revolutions.* (2nd ed.). Chicago: University of Chicago Press, 1970.

Lacoarret, M. et Mme. Ter-Menassian Les universités. In R. Taton (Ed.), *Enseignement et diffusion des sciences en France au XVIIIe siècle.* Paris: Hermann, 1964. Pp. 125–168.

Laissus, Y. Le jardin du roi. In R. Taton (Ed.), *Enseignement et diffusion des sciences en France au XVIIIe siècle.* Paris: Hermann, 1964. Pp. 287–342.

Lakatos, I., & Musgrave, A. (Eds.), *Criticism and the growth of knowledge.* Cambridge: Cambridge University Press, 1970.

Land, K. Principles of path analysis. In E. F. Borgatta & G. W. Bohrnstedt (Ed.), *Sociological methodology: 1969.* San Francisco: Jossey-Bass, 1969. Pp. 3–37.

Langer, B. Pneumatic chemistry, 1772-1789: A resolution of conflict. Unpublished doctoral dissertation, University of Wisconsin, 1972.

Lavoisier, A. *Traité élémentaire de chimie, présenté dans un ordre nouveau et après les découvertes modernes.* Paris, 1789.

Levere, T. H. Relations and rivalry: interactions between Britain and the Netherlands in eighteenth-century science and technology. *History of Science*, 1970, *9*, 42–53.

MacRae, D., Jr. Growth and decay curves in scientific citations. *American Sociological Review*, 1969, *34*, 631–635.

Mayer, J. Portrait d'un Chimiste: Guillaume-François Rouelle (1703-1770). *Revue d'Histoire des Sciences*, 1970, *23*, 305–332.

McCann, H. G. The early career of Claude Louis Berthollet. Unpublished manuscript, Princeton University, 1966.

McCormmach, R. Henry Cavendish: a study of rational empiricism in eighteenth-century natural philosophy. *Isis*, 1969, *60*, 293–306.

McDonald, E. The collaboration of Bucquet and Lavoisier. *Ambix*, 1966, *13*, 74–83.

McKie, D. The scientific periodical from 1665 to 1798. *The Philosophical Magazine*, 1948, Commemoration Number (A. Ferguson, Ed.), 122–132. (a)

McKie, D. Scientific societies to the end of the eighteenth century. *The Philosophical Magazine*, 1948, Commemoration Number (A. Ferguson, Ed.), 133–143. (b)

McKie, D. *Antoine Lavoisier: scientist, economist, social reformer.* New York: Collier Books, 1952.

McKie, D. The *'Observations'* of the Abbé François Rozier (1734–93) – I. *Annals of Science*, 1958, *13*, 73–89.

Meldrum, A. Lavoisier's early work in science, 1763-1771. *Isis*, 1933, *19*, 330–363.

Meldrum, A. Lavoisier's early work in science, 1763-1771. *Isis*, 1934, *20*, 396–425.

Merton, R. K. Science and democratic social structure. In *Social theory and social structure.* Glencoe, Ill.: Free Press, 1949. (Reprinted as Chap. 18 in 1968 edition.)

Merton, R. K. Priorities in scientific discovery: a chapter in the sociology of science. *American Sociological Review*, 1957, *22*, 635–659.

Merton, R. K. Singletons and multiples in scientific discovery. *Proceedings of the American Philosophical Society*, 1961, *105*, 470–486.

Merton, R. K. The ambivalence of scientists. *Bulletin of the Johns Hopkins Hospital*, 1963, *112*, 77–97. (a)

Merton, R. K. Resistance to the systematic study of multiple discoveries in science. *European Journal of Sociology*, 1963, *4*, 237–282. (b)

Merton, R. K. *Social theory and social structure* (Rev ed.). New York: Free Press, 1968.

Merton, R. K. Behavior patterns of scientists. *American Scientist*, 1969, *57*, 1–23.

Merton, R. K. *Science, technology and society in seventeenth century England.* (Originally published as *Osiris* 4: Part Two, 1938.) New York: Harper & Row, 1970.

von Meyer, E. *A history of chemistry from earliest times to the present day* (G. McGowan, trans.) (3rd ed.). London: Macmillan, 1906.

Michaud, J. F., & Michaud, L. G. *Biographie universelle, ancienne et moderne*, 1811–62, *1–55*.

Morrell, J. B. The University of Edinburgh in the late eighteenth century: Its scientific eminence and academic structure. *Isis*, 1971, *62*, 158–171.

Morris, R. J., Jr. Lavoisier on fire and air: the memoir of July 1772. *Isis*, 1969, *60*, 374–380.

Morris, R. J., Jr. Lavoisier and the caloric theory. *British Journal for the History of Science*, 1972, *6*, 1–38.

Multhauf, R. P. On the use of the balance in chemistry. *Proceedings of the American Philosophical Society*, 1962, *106*, 210–218.

Multhauf, R. P. *The origins of chemistry.* London: Oldbourne, 1966.

Musson, A. E., & Robinson, E. Science and industry in the late eighteenth century. *Economic History Review*, 1961, *13*, 222–244.

Nagel, E. *The structure of science.* New York: Harcourt Brace & World, 1961.

Neave, E. W. J. Chemistry in Rozier's journal. I. The journal and its editors. *Annals of Science*, 1950, *6*, 416–421.

Neave, E. W. J. Chemistry in Rozier's journal. II. The phlogiston theory. *Annals of Science*, 1951, *7*, 101–106. (a)

Neave, E. W. J. Chemistry in Rozier's journal. III. Pierre Bayen. *Annals of Science*, 1951, *7*, 144–148. (b)

Neave, E. W. J. Chemistry in Rozier's journal. IV and V. *Annals of Science,* 1951, *7,* 284–299. (c)

Neave, E. W. J. Chemistry in Rozier's journal. VIII and IX. *Annals of Science,* 1952, *8,* 28–45.

Parsons, T. *The social system.* Glencoe, Illinois: The Free Press, 1951.

Partington, J. R. *A short history of chemistry.* London: Macmillan, 1937.

Partington, J. R. Lavoisier's memoir on the composition of nitric acid. *Annals of science,* 1953, *9,* 96–97.

Partington, J. R. Berthollet and the antiphlogistic theory. *Chymia,* 1959, *5,* 130–137.

Partington, J. R. *A history of chemistry.* (Vol. 3). London: Macmillan, 1962.

Perrin, C. E. Prelude to Lavoisier's theory of calcination: some observations on *Mercurius Calcinatus Per Se. Ambix,* 1969, *16,* 140–151.

Perrin, C. E. Early opposition to the phlogiston theory: Two anonymous attacks. *British Journal of the History of Science,* 1970, *5,* 128–144.

Perrin, C. E. Lavoisier, Monge, and the synthesis of water, a case of pure coincidence? *British Journal of the History of Science,* 1973, *6,* 424–428.

Poggendorff, J. C. *Biographisch-literarisches Handwörterbuch zur Geschichte der Exacten Naturwissenschaften.* (Photo reprint, original, 1863.) Amsterdam: Israel, 1965.

Popper, K. *The logic of scientific discovery.* New York: Basic Books, 1959.

de Solla Price, D. J. *Little science, big science.* New York: Columbia University Press, 1963.

de Solla Price, D. J. Citation measures of hard science, soft science, technology, and non-science. In C. E. Nelson & K. Pollock (Eds.), *Communication among scientists and engineers.* Lexington, Massachusetts: Heath, 1970. Pp. 3–22.

Proust, J. *L'encyclopédisme dans le bas-languedoc au XVIIIme siècle.* (Edité par le Centre D'Etudes du XVIIIme Siècle et le Centre d'Études Occitanes, Faculté des Lettres et Sciences Humaines de Montpellier.) Nimes: Imprimerie Barnier, 1968.

Rappaport, R. G.-F. Rouelle: An eighteenth century teacher and chemist. *Chymia,* 1961, *7, 6,* 68–101.

Rappaport, R. Rouelle and Stahl – the phlogistic revolution in France. *Chymia,* 1961, *7,* 73–102.

Rappaport, R. The early disputes between Lavoisier and Monnet, 1777–1781. *British Journal of the History of Science,* 1969, *4,* 233–244.

Robinson, E. The Derby Philosophical Society. *Annals of Science,* 1953, *9,* 359–367.

Robinson, E. Training captains of industry: the education of Matthew Robinson Boulton (1770–1842) and the younger James Watt (1769–1848). *Annals of Science,* 1954, *10,* 301–313.

Robinson, E. Thomas Beddoes, M. D., and the reform of science teaching in Oxford. *Annals of Science,* 1955, *11,* 137–144.

Robinson, E. The Lunar Society and the improvement of scientific instruments: I. *Annals of Science,* 1957, *12,* 296–304.

Robinson, E. The Lunar Society and the improvement of scientific instruments: II. *Annals of Science,* 1958, *13,* 1–8.

Russell-Wood, J. The scientific work of William Brownrigg, M. D., F. R. S. (1711-1800) II. *Annals of Science,* 1951, *7,* 77–94.

Scheffler, I. *Science and subjectivity.* Indianapolis: Bobbs–Merrill, 1967.

Schoenberg, R. Strategies for meaningful comparison. In H.L. Costner (Ed.), *Sociological methodology: 1972.* San Francisco: Jossey-Bass, 1972. Pp. 1–35.

Schofield, R. E. Membership of the Lunar Society of Birmingham. *Annals of Science,* 1957, *12,* 118–136. (a)

Schofield, R. E. The industrial orientation of science in the Lunar Society of Birmingham. *Isis,* 1957, *48,* 408–415. (b)

Schofield, R. E. Josiah Wedgewood, industrial chemist. *Chymia,* 1959, *5,* 180–192.

Schofield, R. E. Still more on the water controversy. *Chymia*, 1963, *9*, 71–76. (a)

Schofield, R. E. *The Lunar Society of Birmingham: A social history of provincial science and industry in eighteenth-century England.* Oxford: Clarendon Press, 1963. (b)

Schofield, R. E. Joseph Priestley, the theory of oxidation and the nature of matter. *Journal of the History of Ideas*, 1964, *25*, 285–294.

Schofield, R. E. The Lunar Society of Birmingham: a bicentenary appraisal. *Notes and Records of the Royal Philosophical Society, London*, 1966, *21*, 144–161.

Schofield, R. E. Joseph Priestley, natural philosopher. *Ambix*, 1967, *14*, 1–15.

Schofield, R. E. *Mechanism and materialism: British natural philosophy in an age of reason.* Princeton, N.J.: Princeton University Press, 1970.

Scudder, S. H. *Catalogue of scientific serials: 1633–1876.* Cambridge, Massachusetts: Harvard University Press, 1879. (Reprinted by Kraus Reprint Corp., New York, 1965.)

Senac, J. B. *Nouveau cours de chymie, suivant les principles de Newton & de Stahl,* 2 vols. Paris, 1723.

Siegfried, R. Lavoisier's view of the gaseous state and its early application to pneumatic chemistry. *Isis*, 1972, *63*, 59–78.

Siegfried, R., & Dobbs, B. J. Composition: A neglected aspect of the chemical revolution. *Annals of Science*, 1968, *24*, 275–293.

Sivin, N. William Lewis (1708–1781) as chemist. *Chymia*, 1962, *8*, 63–88.

Smeaton, W. A. The contributions of P. -J. Macquer, T. O. Bergman, and L. B. Guyton de Morveau to the reform of chemical nomenclature. *Annals of Science*, 1954, *10*, 87–106. (a)

Smeaton, W. A. The early history of laboratory instruction in chemistry at the Ecole Polytechnique, Paris, and elsewhere. *Annals of Science*, 1954, *10*, 224–240. (b)

Smeaton, W. A. Lavoisier's membership of the Société Royale de Médcine. *Annals of Science*, 1957, *12*, 228–240. (a)

Smeaton, W. A. Lavoisier's membership of the Société Royale d'Agriculture and the Comité d'Agriculture. *Annals of Science*, 1957, *12*, 267–277. (b)

Smeaton, W. A. *L'Avant-Coureur.* The journal in which some of Lavoisier's earliest research was reported. *Annals of Science*, 1959, *13*, 219–234.

Smeaton, W. A. Guyton de Morveau's course of chemistry in the Dijon Academy. *Ambix*, 1961, *9*, 53–69.

Smeaton, W. A. *Fourcroy, chemist and revolutionary: 1755–1809.* Cambridge, England: Heffer, 1962.

Smeaton, W. A. Louis Bernard Guyton de Morveau, F. R. S. (1737–1816) and his relations with British scientists. *Notes and Records of the Royal Philosophical Society, London*, 1967, *22*, 113–129.

Smeaton, W. A. New light on Lavoisier: The research of the last ten years. *History of Science*, 1970, *9*, 51–69.

Smith, B. M. D., & Moilliet, J. L. James Keir of the Lunar Society. *Notes and Records of the Royal Philosophical Society, London*, 1967, *22*, 144–154.

Stahl, G. E. *Fundamenta chymiae dogmaticae & experimentalis.* Nuremberg, 1723. (Trans. by P. Shaw as *Philosophical principles of universal chemistry. Drawn from the Collegium Jenese of G. E. Stahl.* London, 1730.)

Storer, N. W. *The social system of science.* New York: Holt, Rinehart, & Winston, 1966.

Taton, R. *Enseignement et diffusion des sciences en France au XVIIIe siècle.* Histoire de la Pensée XI, Ecole Pratique des Hautes Etudes, Sorbonne. Paris: Hermann, 1964.

Taylor, F. S. The teaching of the physical sciences at the end of the eighteenth century. *The Philosophical Magazine*, 1948, Commemoration Number (A. Ferguson, Ed.), 144–164.

Thackray, A. W. Science and technology in the industrial revolution. *History of Science*, 1970, *9*, 76–89. (a)

Thackray, A. W. *Atoms and powers.* Cambridge, Massachusetts: Harvard University Press, 1970, (b)

Torlais, J. Le collège royal. In R. Taton (Ed.), *Enseignement et diffusion des sciences en France au XVIIIe siècle.* Paris: Hermann, 1964.

Toulmin, S. Crucial experiments: Priestley and Lavoisier. *Journal of the History of Ideas,* 1957, *18,* 205–220.

Trengove, L. Chemistry at the Royal Society of London in the eighteenth century — I. *Annals of Science,* 1965, *19,* 183–283.

Tukey, J. W., & Wilk, M. B. Data analysis and research design. In E. R. Tufte (Ed.), *The quantitative analysis of social problems.* Reading, Massachusetts: Addison–Wesley, 1970. Pp. 370–390.

Wightman, W. P. D. William Cullen and the teaching of chemistry. *Annals of Science,* 1955, *11,* 154–165.

Wightman, W. P. D. William Cullen and the teaching of chemistry — II. *Annals of Science,* 1957, *12,* 192–205.

Williams, L. P. The politics of science in the French Revolution. In M. Clagett (Ed.), *Critical problems in the history of science.* Madison, Wisconsin: University of Wisconsin Press, 1959. Pp. 291–308.

Wolf, A. In D. McKie (Ed.), *History of science, technology, and philosophy in the 18th century* (2nd ed.). London: Allen & Unwin, 1952.

Wright, S. Correlation and causation. *Journal of Agricultural Research,* 1921, *20,* 557–585.

Wright, S. The method of path coefficients. *Annals of Mathematical Statistics,* 1934, *5,* 161–215.

Wright, S. The interpretation of multivariate systems. In O. Kempthorne et al. (Eds.), *Statistics and mathematics in biology.* Ames, Iowa: Iowa State College Press, 1954. Pp. 11–33.

Wright, S. Path coefficients and path regressions: Alternative or complementary concepts? *Biometrics,* 1960, *16,* 189–202.

Znaniecki, F. *The social role of the man of knowledge.* New York: Columbia University Press, 1940.

Zuckerman, H., & Merton, R. Patterns of evaluation in science: institutionalization, structure and functions of the referee system. *Minerva,* 1971, *9,* 66–100.

Author Index

Subject Index